AF478589

DEEP MILLIMETER SURVEYS:
IMPLICATIONS FOR GALAXY FORMATION AND EVOLUTION

DEEP MILLIMETER SURVEYS:
IMPLICATIONS FOR GALAXY FORMATION AND EVOLUTION

19 – 21 June, 2000

University of Massachusetts, Amherst, USA

Editors

James D. Lowenthal
University of Massachusetts, Amherst, USA

David H. Hughes
Instituto Nacional de Astrofísica,
Óptica, y Electrónica, Puebla, Mexico

Published by

World Scientific Publishing Co. Pte. Ltd.

P O Box 128, Farrer Road, Singapore 912805

USA office: Suite 1B, 1060 Main Street, River Edge, NJ 07661

UK office: 57 Shelton Street, Covent Garden, London WC2H 9HE

British Library Cataloguing-in-Publication Data
A catalogue record for this book is available from the British Library.

Cover design: Confusion-limited simulation of an extragalactic 1.1mm survey with the Large Millimetre Telescope, courtesy of Enrique Gaztañaga and David H. Hughes.

DEEP MILLIMETER SURVEYS
Implications for Galaxy Formation and Evolution

ISBN 981-02-4465-7

Printed in Singapore by Uto-Print

Preface

The UMass/INAOE Conference on "Deep Millimeter Surveys: Implications For Galaxy Formation And Evolution" gathered together for the first time an international field of about 70 experts at the University of Massachusetts campus in Amherst from 19-21 June 2000 to compare notes, present new data, and discuss the implications of the new results and models for our understanding of galaxy formation and evolution at sub-mm and mm-wavelengths — a new window opened only in the last three years.

The three days of very stimulating talks and discussion focussed on exciting new results from the SCUBA detector array; on the expected performance of new facilities coming on line in the next few years, such as the Large Millimeter Telescope (currently under construction in Mexico in a major collaboration between UMass and INAOE, the conference sponsors) and the Atacama Millimeter Array; and on how to use known local and distant galaxies – mostly starbursts – to interpret the new sub-mm and millimeter-wave results. Both observations and theory were evident in abundance, and practitioners of both had ample opportunity to compare notes over coffee and dinner, including a New England-style lobster bake for the Conference Dinner.

An excellent opening talk by Simon Lilly and a provocative summary by Roberto Terlevich helped to keep the big picture of galaxy formation and evolution in mind while at the same time maintaining the details in focus.

We were fortunate to have UMass Conference Services in charge of the logistics, the food, and the registration, and thanks are due to them for making everything go as smoothly as it did. Graduate students from UMass also provided valuable help setting up posters, staffing the computers and projecters, and taking care of lighting.

Finally, we are indebted to our sponsors for providing the financial support required to put the conference on in the first place: The LMT US Project Office; the Instituto Nacional de Astrofísica, Óptica, y Electrónica (INAOE); the UMass Office of Vice Chancellor for Research; the UMass College of Natural Science and Mathematics; the UMass Department of Astronomy; and Composite Optics, Incorporated.

A second joint UMass/INAOE meeting on mm-wave extragalactic science will be held at INAOE headquarters in Tonantzintla, Mexico in the summer of 2002, and we look forward to seeing there many of the same faces we saw in Amherst in 2000.

Deep Millimeter Surveys
Conference Organizing Committee

UMass   **INAOE**

UMass	INAOE
James D. Lowenthal (Co-chair)	David H. Hughes (Co-chair)
William Irvine	Elena Terlevich
Peter Schloerb	Roberto J. Terlevich
Steve Schneider	Enrique Gatzañaga
Ron Snell	Alberto Carramiñana
Judy Young	Luis Carrasco

Contents

1. Continuum Sub-mm/mm Surveys

THE NATURE OF FAINT SUBMILLIMETER GALAXIES

IAN SMAIL

Department of Physics, University of Durham, South Road, Durham DH1 3LE
E-mail: ian.smail@durham.ac.uk

ROB IVISON

Department of Physics & Astronomy, University College London, Gower Street,
London WC1E 6BT E-mail: rji@star.ucl.ac.uk

ANDREW BLAIN

Institute of Astronomy, Madingley Road, Cambridge CB3 0HA
E-mail: awb@ast.cam.ac.uk

JEAN-PAUL KNEIB

Observatoire de Toulouse, 14 avenue E. Belin, 31400 Toulouse, France
E-mail: jean-paul.kneib@ast.obs-mip.fr

We summarise the main results on the faint submillimeter (submm) galaxy population that have come from the SCUBA Cluster Lens Survey. We detail our current understanding of the characteristics of these submm-selected galaxies across wavebands from X-rays to radio. After presenting the main observational properties of this population we conclude by discussing the nature of these distant, ultraluminous infrared galaxies and their relationship to other high-redshift populations.

1 Introduction

The results of the highly successful far-infrared (FIR) survey undertaken by *IRAS* led to a wide-spread realisation of the ubiquity and importance of highly-obscured star-forming and active galaxies in the local universe. More recent work in the FIR and submm wavebands has produced a similar revolution in our understanding of obscured galaxies in the distant universe. These observations employ the *COBE* and *ISO* satellites and the Sub-millimeter Common User Bolometer Array (SCUBA[1]) on the 15-m JCMT[a] and have shown that obscured galaxies contribute a substantial fraction of the total emitted radiation at high redshifts.

The rough equivalence of the energy density in the optical background and that detected in the FIR/submm by *COBE* suggests that, averaged over

---

[a]The JCMT is operated by the Joint Astronomy Centre on behalf of the United Kingdom Particle Physics and Astronomy Research Council (PPARC), the Netherlands Organisation for Scientific Research, and the National Research Council of Canada.

all epochs, approximately half of the total radiation in the universe came from obscured sources (either stars or AGN). Clearly including this class of galaxies in models of galaxy evolution is critical to obtaining a complete understanding of the formation and evolution of galaxies.

As we will show in the next section, the bulk of the emission in the FIR/submm comes from a relatively small population of extremely luminous, dusty galaxies. These galaxies lie at high redshifts, $z \gtrsim 1$–4, are both massive and gas-rich and they may dominate the total star formation in the universe at these early epochs. The analogs of this population in the local universe are the Ultra-Luminous Infrared Galaxies (ULIRGs) uncovered by *IRAS*. As a benchmark for the following discussion we note that a ULIRG similar to Arp 220 with a far-infrared luminosity of $L_{FIR} \sim 3 \times 10^{12} L_\odot$ and a star-formation rate (SFR) of $\sim 300 \, M_\odot \, \mathrm{yr}^{-1}$ would have an 850-μm flux density of $\gtrsim 3 \, \mathrm{mJy}$ out to $z \sim 10$ in a spatially flat Universe.[b]

2 Submm number counts and the FIR background

The advent of sensitive submm imaging with SCUBA has allowed a number of groups to undertake 'blank'-field surveys for faint submm galaxies. Results on the number density of sources in blank fields as a function of 850-μm flux density have been published by three groups.[2,3,4] Unfortunately, due to the modest resolution of the SCUBA maps, $15''$ FWHM, these surveys are confusion limited at $\sim 2 \, \mathrm{mJy}$ and resolve $\sim 50\%$ of the *COBE* background.

Our collaboration has taken a complementary approach to these 'blank' field surveys by using massive gravitational cluster lenses to increase the sensitivity and resolution of SCUBA. This survey covers seven lensing clusters at $z = 0.19$–0.41[5,6,7] (new results from two similar lensing surveys with SCUBA were also reported at this meeting[8,9]). Our analysis uses well-constrained lens models to correct the observed source fluxes for lens amplification.[7] For the median source amplification, $\sim 2.5\times$, this survey covers an area of the source plane equivalent to 15 arcmin^2 at a 3σ flux limit of $\sim 2 \, \mathrm{mJy}$. The amplification also results in a factor of two finer beam size at this depth so that these counts have a fainter confusion limit than the blank field observations. The surface density of sources at the limit of our survey (0.5 mJy) is $\sim 8 \, \mathrm{arcmin}^{-2}$ (the equivalent density of 'normal' field galaxies is reached at $I \sim 23.5$) and we resolve $\sim 100\%$ of the *COBE* background.[7] We conclude that the majority of the extragalactic submm background arises in a relatively sparse population of extremely luminous galaxies, $L \gtrsim 10^{12} L_\odot$.

---

[b]We assume $q_o = 0.5$ and $h_{100} = 0.5$ unless otherwise stated.

3 Optical counterparts to submm galaxies

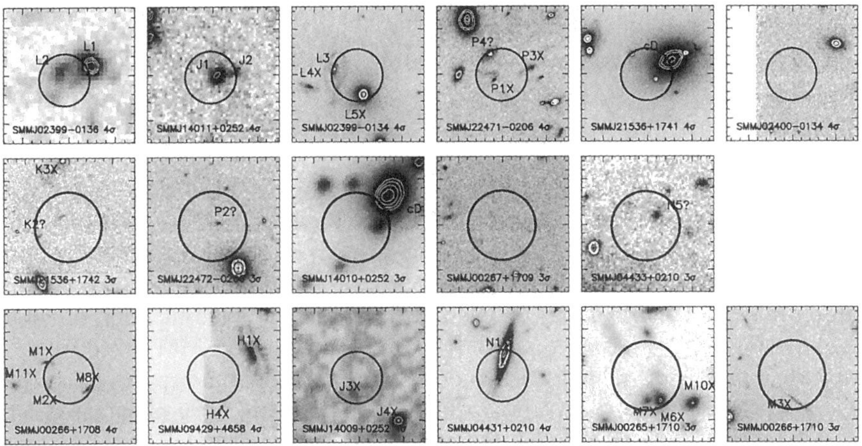

Figure 1. $15'' \times 15''$ images of the 17 sub-mm sources in our full sample. These are ordered from the upper-left on the basis of the reliability of their proposed optical counterparts – with the first two rows containing those sources that were probably correctly identified in the optical imaging.[21,25] Possible counterparts are identified with '?'; those that are unlikely to be the correct counterpart are marked with an 'X'. The bottom row shows those submm source where subsequent near-infrared and radio analysis has brought into question the identification of the proposed optical counterparts.

The first attempts at identifying counterparts to the submm sources in our survey concentrated on exploiting the high-quality, archival *Hubble Space Telescope WFPC2* imaging that existed for the majority of these fields.[6] These frames typically reach limits of $I \sim 27$ on the background source-plane and we felt confident of identifying a large fraction of the submm galaxies (reflecting the optical bias of the author). Figure 1 shows the identifications from Smail et al. (1998)[6] but ordered in terms of our current understanding of the reliability of the proposed counterparts (based on the near-infrared and radio follow-up discussed in the next section).

If we remove the two central cluster galaxies[10] from our sample (SMM J21536+1741 and SMM J14010+0252) – we conclude that only about half of the original identifications that were proposed are likely to be correct. Indeed, of all the submm galaxies detected in all the SCUBA surveys, only three have confirmed counterparts: SMM J02399−0136,[11] SMM J14011+0252[12] and SMM J02399−0134 (Fig. 1), where the confirmation comes from identification of CO emission at the same redshift as the proposed counterpart.[13,14,15] For a handful of other sources, their continuum detections with millimeter interferometers and the improved astrometry that these provide have been used to identify likely counterparts.[16,17] Equally, identifications of counterparts have been claimed when a galaxy with sufficiently unusual characteristics is found within the submm error box (e.g. an ERO[18]), but so far none of these have been confirmed in CO.

The relatively low identification rate, even with the high-quality optical imaging we have available, results partly from the fact that source confusion with the large SCUBA beam is compounded by clustering, potentially leading to deviations of the order of the beam-size.[19] Indeed the clearest route to identifying counterparts is usually to overlay the raw SCUBA map onto the optical image and to use the information present in the shape of the submm emission to search for possible merging of sources. We believe that source confusion may have contributed to the ambiguities in the identification of counterparts for SMM J21536+1742, SMM J22471−0206 and SMM J02400−0134. However, the main difficulty with identifying optical counterparts arises because the majority of the submm population are *extremely* faint at optical wavelengths: at least half and perhaps up to 75% of galaxies in Fig. 1 have optically faint counterparts, $I \gg 25$–27 (i.e. Classes 0 and 1[20]). Removing the known AGN from the sample, the proportion of starburst-powered submm galaxies that have optically faint counterparts rises to 90%. We now briefly discuss some of the observations that have led us to this conclusion.

4 Near-infrared and radio counterparts to submm galaxies

The first indication that things might not be as clear-cut as the proposed optical identifications suggested came from K-band imaging of our SCUBA fields using UKIRT.[c] These images reach $K \sim 21$ in the source plane and uncovered new candidates (H5 and N4 in Fig. 2) within the submm error-boxes of two of the submm sources.[18] These galaxies are undetected in very deep *HST* and Keck R- and I-band imaging ($I \gtrsim 27$ corrected for lens amplification), but are relatively bright in the K-band. Their colours, $(I - K) \gtrsim 6.0$ and $\gtrsim 6.8$ for H5 and N4 respectively, place them firmly in the rare class of Extremely Red Objects (EROs).[18] The surface density of ERO-submm galaxies in our survey means that they account for around half of all ERO's – providing an important link between these two populations. Moreover, the extreme submm to optical ratios of these galaxies, $L_{FIR}/L_B \gtrsim 300$ (assuming they lie at $z \sim 2.5$–3, as suggested by fits to their SEDs), combined with their high inferred star formation rates, underlines the difficulties faced when attempting

---

[c]UKIRT is operated by the Joint Astronomy Centre on behalf of PPARC.

to use UV-selected samples to audit star formation at $z \gg 1$.

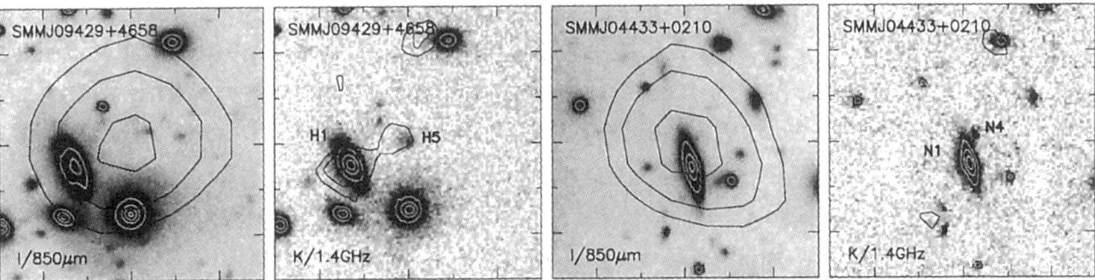

Figure 2. ERO counterparts to two submm sources in our survey.[18] The two panels for each source show deep I- and K-band images with the 850 μm and 1.4 GHz maps overlayed. The faintest sources visible in the I-band exposure have observed magnitudes of $I \sim 25.5$–26.0 and $K \sim 20.5$. The original counterparts proposed for the submm sources are marked on the K-band images, as well as the new ERO candidates, H5 for SMM J09429+4658 and N4 for SMM J04433+0210. Each panel is 30″ square and is centred on the 850-μm position.

The next phase of the identification process used *very* deep 1.4 GHz maps of our fields from the VLA.[21] These maps allow us to identify radio counterparts or place stringent limits ($\lesssim 20\mu$Jy in the source plane) on the radio flux of the submm galaxies. Moreover, the recent work on submm-to-radio spectral indices for distant star forming galaxies[22,23,24] has provided us with a useful tool for estimating the redshift distribution of a *complete* submm-selected sample.[21] We can compare the redshift limits derived in this manner with the spectroscopic redshifts for individual candidate optical counterparts[25] and hence determine the reliability of these proposed counterparts as shown in Fig. 1. On the basis of this analysis we conclude that the submm population brighter than ~ 1 mJy has a median redshift of *at least* $<z> \sim 2$, and using the available spectral index models, probably $<z> \sim 2.5$–3.[21] The high median redshift means that the submm population, if predominately powered by starbursts (rather than AGN), contributes a substantial fraction of the total star formation density at high redshifts.

Our on-going program of observations of this sample of submm galaxies, using near-infrared imaging and spectroscopy and millimeter interferometry, is reported elsewhere in this volume.[26]

5   The submm population at other wavelengths

To understand the role of the submm galaxies in galaxy formation and evolution we first have to determine the relative contributions to their extreme luminosities from AGN and starburst-powered emission. Hard X-ray observations can uncover the presence of a dominant AGN power source in even highly obscured sources, and hence provide a simple estimate of the AGN fraction in the submm population.[27,28] Unfortunately, even for high-redshift sources, *Chandra* provides relatively poor hard X-ray response, and hence to-date all the X-ray studies of submm galaxies[29,30] have achieved is to confirm the presence of modestly-obscured AGN in those galaxies in which optical spectroscopy had already identified AGN emission.[11,25] These galaxies account for 10–20% of the submm population, in-line with estimates from the models of the X-ray background.[27,28] Much more interesting limits on the fraction of highly-obscured AGN in this population will come from *Newton* observations. Equally useful constraints on the fraction of obscured AGN and the breakdown of AGN- versus starburst-powered emission in individual galaxies will come from high-resolution, mid-infrared observations (using OSCIR or Michelle on Gemini) that are sensitive to emission from hot-dust and PAHs. However, we conclude that, on present evidence, it seems safe to assume that the majority of the emission from the submm population is powered by star formation.[31]

6 The relationship between FIR- and UV-selected galaxies

Finally, we briefly discuss the relationship of submm galaxies to other classes of high-redshift sources. Recently it has been claimed that UV- and optically-selected galaxy samples at $z \sim 0$–3 can completely account for the extragalactic background, not only in the optical/UV, but also in the far-infrared and submm.[32] This calculation relies upon the correction of the UV luminosities of these samples for the effects of dust extinction, using a correlation between reddening (measured through the spectral slope in the UV) and UV luminosity from observations of normal galaxies at low redshifts. Unfortunately, as was graphically illustrated at this conference,[33] this correlation doesn't hold (even locally) for galaxies with bolometric luminosities typical of the population we have found at higher redshifts that dominate the submm counts and produce the bulk of the *COBE* background. This clearly calls into question the reliability of this type of calculation.

The argument that UV-selected samples can provide a complete picture is supported using optical/UV and submm observations of an example of a 'representative' submm galaxy, SMM J14011+0252 (Fig. 3).[12] However, this galaxy is far from typical of the majority of the submm population – which would be better described using the ERO, HR 10, as an archetype.[34] Moreover, even within the SMM J14011+0252 system it appears from interferometry observations that the dominant bolometric component resides, not in the UV-bright companion J2, but in the much redder galaxy J1. Using a UV-weighted spectral slope (which has a substantial contribution from J2) in an attempt to predict the unobscured star formation rate (which mostly occurs in J1) in this system seems questionable – why should the UV properties of J2 have any bearing on the degree of obscuration and star formation in J1?

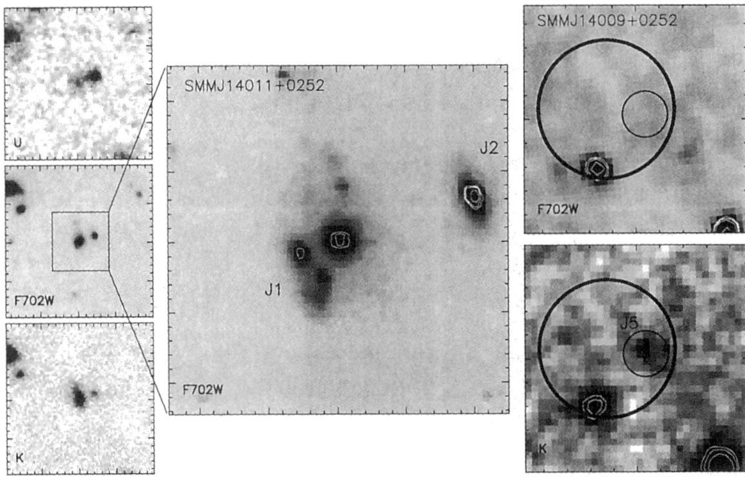

Figure 3. These panels show two of the submm galaxies detected through the lensing cluster A 1835.[12] The left-hand panels illustrate the morphology of the two components, J1/J2, of the $z = 2.55$ starburst galaxy SMM J14011+0252, from the U- to K-bands. The central panel gives a zoomed view of the *WFPC2* R-band image of this galaxy, demonstrating the complex, merger-like morphology of the bolometrically-dominant component J1. The right-hand panels show R- and K-band views of a more typical submm galaxy, SMM J14009+0252, for which a very faint near-infrared companion, J5, has been identified.[12] This galaxy is clearly undetected in the deep *HST* R-band exposure. The major tickmarks are every $5''$, except for the central panel where they are every $1''$.

Instead, if we adopt HR 10 as a template for a typical submm galaxy then we find that: *as HR10 has a far higher ratio of far-IR to far-UV luminosity than the other galaxies considered it would not be included in UV-selected surveys despite its large star formation rate.*[32] We conclude therefore that UV light is an inappropriate tracer of star formation in the highly-obscured galaxies that contain a large fraction of the star formation in the high redshift universe. Further study of this remarkable population is essential for a complete understanding of the formation of massive galaxies.[31,20]

Acknowledgments

We thank our collaborator on the *HST* lensing project, Harald Ebeling, for allowing us to present the *WFPC2* image of A 1835. IRS acknowledges travel support to attend this conference. The research presented here was supported by the Royal Society [IRS], PPARC [RJI], the Raymond and Beverly Sackler Foundations [AWB] and the CNRS [JPK].

References

1. Holland, W.S., et al., MNRAS **303**, 659 (1999).
2. Hughes, D.H., et al., Nature **394**, 241 (1998).
3. Eales, S., et al., ApJ **515**, 518 (1999).
4. Barger, A.J., et al., Nature **394**, 248 (1998).
5. Smail, I., Ivison, R.J., Blain, A.W., ApJ **490**, L5 (1997).
6. Smail, I., Ivison, R.J., Blain, A.W., Kneib, J.-P., ApJ **507**, L21 (1998).
7. Blain, A.W., Kneib, J.-P., Ivison, R.J., Smail, I., ApJ **512**, L87 (1999).
8. van der Werf, P., et al., these proceedings, (2001)
9. Scott, D., et al., these proceedings, (2001)
10. Edge, A.C., et al., MNRAS **306**, 599 (1999).
11. Ivison, R.J., et al., MNRAS **298**, 583 (1998).
12. Ivison, R.J., et al., MNRAS **315**, 209 (2000).
13. Frayer, D.T., et al., ApJ **506**, L7 (1998).
14. Frayer, D.T., et al., ApJ **514**, L13 (1999).
15. Kneib, J.P., et al., A&A, in prep (2001)
16. Gear, W.,K., et al., MNRAS **316**, 51 (2000)
17. Bertoldi, F., et al., A&A **360**, 92 (2000)
18. Smail, I., et al., MNRAS **308**, 1061 (1999).
19. Hogg, D.W., AJ , in press (2000)
20. Ivison, R.J., et al., ApJ , in press (2001).
21. Smail, I., et al., ApJ **528**, 612 (2000).
22. Carilli, C.L., Yun, M.S., ApJ **513**, L13 (1999).
23. Carilli, C.L., Yun, M.S., ApJ **530**, 618 (2000).
24. Blain, A.W., MNRAS **309**, 955 (1999).
25. Barger, A.J., et al., AJ **117**, 2656 (1999).
26. Frayer, D.T., these proceedings, (2001).
27. Almaini, O., Lawrence, A., Boyle, B.J., MNRAS **305**, 59 (1999)
28. Gunn. K.F, Shanks, T., MNRAS, in press (2000)
29. Fabian, A.C., et al., MNRAS **315**, 8 (2000).
30. Bautz, M.W., et al., ApJ, in press (2000)
31. Blain, A.W., et al., MNRAS **302**, 632 (1999).
32. Adelberger, K.L., Steidel, C.C., ApJ, submitted (2000).
33. Sanders, D., these proceedings, (2001).
34. Dey, A., et al., ApJ **519**, 610 (1999)

SUB-MM CLUES TO ELLIPTICAL GALAXY FORMATION

JAMES S. DUNLOP

Institute for Astronomy, Royal Observatory, Edinburgh EH9 3HJ, UK

There is growing evidence that, at the $S_{850\mu m} < 1\,\mathrm{mJy}$ level, the sub-mm galaxy population (and hence a potentially significant fraction of the sub-mm background) is associated with the star-forming Lyman-break population already detected at optical wavelengths. However, the implied star-formation rates in such objects (typically $3 - 30\,\mathrm{M_\odot yr^{-1}}$) fall one or two orders of magnitude short of the level of star-forming activity required to produce the most massive elliptical galaxies on a timescale $\sim 1\,\mathrm{Gyr}$. If a significant fraction of massive ellipticals did form the bulk of their stars in short-lived massive starbursts at high redshift, then they should presumably be found among the brighter, $S_{850\mu m} \simeq 10\,\mathrm{mJy}$ sub-mm sources which are undoubtedly *not* part of the Lyman-break population. A first powerful clue that this is indeed the case comes from our major SCUBA survey of radio galaxies, which indicates that massive dust-enshrouded star-formation in at least this subset of massive ellipticals is largely confined to $z > 2.5$, with a mean redshift $z \simeq 3.5$. While radio selection always raises concerns about bias, I argue that our current knowledge of the brightest ($S_{850\mu m} \simeq 10\,\mathrm{mJy}$) sub-mm sources detected in un-biassed SCUBA imaging surveys indicates that they are also largely confined to this same high-z regime. Consequently, while the most recent number counts imply such extreme sources can contribute only $5 - 10\%$ of the sub-mm background, their comoving number density (in the redshift band $3 < z < 5$) is $\simeq 1 - 2 \times 10^{-5}\mathrm{Mpc^{-3}}$, sufficient to account for the formation of *all* ellipticals of comparable mass to radio galaxies ($\geq 4L^\star$) in the present-day universe.

1 Introduction

The formation mechanism of elliptical galaxies remains a fundamental and controversial issue in cosmology. In current models of galaxy formation dominated by cold dark matter (CDM), elliptical galaxies arise from the merging at low redshift of intermediate-mass discs[2], and some recent data have been interpreted as supportive of the implied gradual formation of massive ellipticals at relatively low redshift[15]. However, the validity of these analyses has recently been questioned[25,14], and other observational evidence at low/moderate redshift[4,19,3], continues to favour a picture in which at least some massive ellipticals formed the bulk of their stars in a short-lived ($\leq 1\,\mathrm{Gyr}$) massive starburst at high redshift ($z > 3$)[14]. While this high-redshift star-formation scenario might only apply to a subset of ellipticals[18], it can be argued that it applies to massive ellipticals in general[22]. Moreover, the clarification of the link between black-hole and spheroid mass[17] suggests that the hosts of AGN may be more representative of spheroids in general than previously supposed.

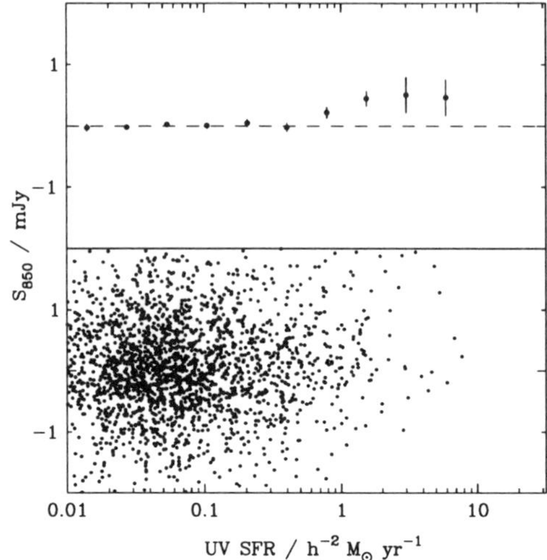

Figure 1. A plot of observed 850-μm flux density against UV SFR for known optically-selected starburst galaxies in the HDF. The bottom panel shows the raw values, while the top panel shows the mean values with standard errors. A statistically significant sub-mm detection is obtained for star-forming galaxies with $SFR > 1\,h^{-2}\mathrm{M_\odot yr^{-1}}$

2 Observing Spheroid Formation - Optical versus Sub-mm

It has been argued that the formation of present-day galactic bulges/spheroids has already been observed at optical wavelengths, through the discovery of the Lyman-break population at $z \simeq 2-4$ [24]. However, even if substantial corrections are made to correct for the effects of dust, the inferred star-formation rates in these objects are relatively modest (typically $3-30\,\mathrm{M_\odot yr^{-1}}$)[21]. Even luminous Lyman-break objects thus appear to fall over an order of magnitude short of the high star-formation rates ($\simeq 1000\,\mathrm{M_\odot yr^{-1}}$) required to construct the stellar populations of the most massive ellipticals on a timescale $\leq 1\,\mathrm{Gyr}$.

Direct confirmation of this comes not only from the difficulty experienced in detecting individual, unlensed Lyman-break objects with SCUBA [5], but also from the achievement of a statistical detection of the bright end of the Lyman-break population in the deep SCUBA image of the HDF [20]. The basic evidence for this result is shown in Fig.1; Lyman-break galaxies with raw (uncorrected, UV-derived) star-formation rates $\simeq 1\,h^{-2}\mathrm{M_\odot yr^{-1}}$ are detected, statistically, in the SCUBA image at flux densities of $S_{850\mu\mathrm{m}} \simeq 0.2$ mJy. This number implies that the ratio of hidden-to-visible star-formation in these

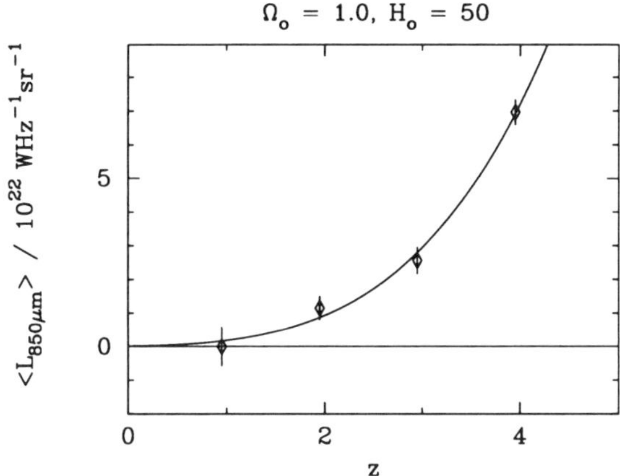

Figure 2. The rapid growth of average sub-millimetre luminosity with increasing redshift found for powerful radio galaxies spanning the redshift range $0.5 < z < 4.5$. The datapoints are the result of SCUBA observations of a sample of $\simeq 50$ radio galaxies, placed in redshift bins of unit width. The curve has the form $L \propto (1 + z)^4$, and serves to illustrate the dramatic nature of the increase in characteristic sub-mm luminosity, particularly beyond $z \simeq 2.5$. This result suggests that the epoch of maximum star-formation in the massive elliptical hosts of radio sources lies at $z \simeq 3 - 4$.

objects is 5 ± 1.5 (consistent both with dust-screen models[21] and with models which assume the first '15 Myrs-worth' of star-formation in giant molecular clouds is always essentially invisible at UV wavelengths [14]). It also implies that the Lyman-break population is probably responsible for a significant, albeit highly-uncertain fraction of the sub-mm background ($\sim 25\%$).

This independent check on the hidden:visible star-formation ratio in Lyman-break galaxies confirms that optical surveys rarely reveal galaxies forming stars at the level of $100 - 1000\,\mathrm{M_\odot yr^{-1}}$. One explanation for this is that the formation of the stellar populations of massive ellipticals is too widely distributed in space and/or time to be identified with a very violent event. Observationally, however, it remains possible that such massive starbursts are prevalent at high redshift, but are either too dust-enshrouded, or at too extreme a redshift to be detected in existing optical-UV drop-out surveys.

In fact, since the first sub-mm detections of the $z \simeq 4$ radio galaxies 4C41.17 and 8C1435+635, it has been clear that massive dust-enshrouded high-z starbursts do at least exist[7,12]. These galaxies display exactly the properties expected of a young massive elliptical, with inferred dust-enshrouded star-formation rates $\simeq 1000\,\mathrm{M_\odot yr^{-1}}$. However, the relevance of such extreme sources to the general elliptical population remains uncertain.

3 Sub-mm Studies of Radio Galaxies out to $z \simeq 4$

3.1 SCUBA Photometry of Radio Galaxies

These pioneering sub-mm observations have therefore raised the important issue of whether all luminous sub-mm sources are confined to extreme redshift ($z \simeq 3 - 5$). A powerful clue that this may be true comes from our major SCUBA study[1] of radio galaxies spanning the redshift range $0 < z < 5$. Such a survey is potentially biased, being based on an (arguably) special subset of massive ellipticals. However, it does offer the advantages of ready-made redshift information and accurately known positions (allowing SCUBA to be used in its most sensitive photometry mode). These advantages have enabled us to deduce that the typical level of dust-enshrouded star-formation in radio galaxies grows rapidly beyond $z \simeq 2$, continuing to rise out to $z \simeq 4$ (Fig.2).

How biased might this result be? Interestingly the median redshift of the submm detections in the radio-galaxy sample is $z \simeq 3$, consistent with that derived for the (cluster-lensed) field population[1]. However, as shown in Fig.2, the average sub-mm luminosity continues to rise beyond $z \simeq 3$. This is because 4 of the 5 most luminous sub-mm sources ($S_{850\mu m} > 5$ mJy) in the sample lie at $z > 3$. Could this be true for luminous sub-mm sources in general?

3.2 SCUBA Imaging of Radio Galaxies

Already it is clear that large sub-mm luminosities at $z \simeq 3 - 4$ are not simply confined to the host galaxies of AGN. New, deep, SCUBA *imaging* of the regions around several of the above-mentioned radio galaxies has revealed even more submm-luminous companion sources, at the same redshift[13]. The implication is that high-z radio galaxies act as signposts towards young clusters in which much of the eventual stellar content of the cluster ellipticals is forming in massive dust-enshrouded starbursts ($\simeq 1000 \, M_\odot \mathrm{yr}^{-1}$). This is consistent with the apparent age and coevality of present-day cluster ellipticals. Moreover, the faintness of the possible optical IDs of these very luminous sub-mm sources indicates that this process is basically invisible at optical wavelengths.

4 Bright Sources from Un-biased Sub-mm Surveys

With the advent of sensitive sub-mm imaging with SCUBA, un-biased sub-mm surveys with the potential to properly quantify the prevalence of massive dust-

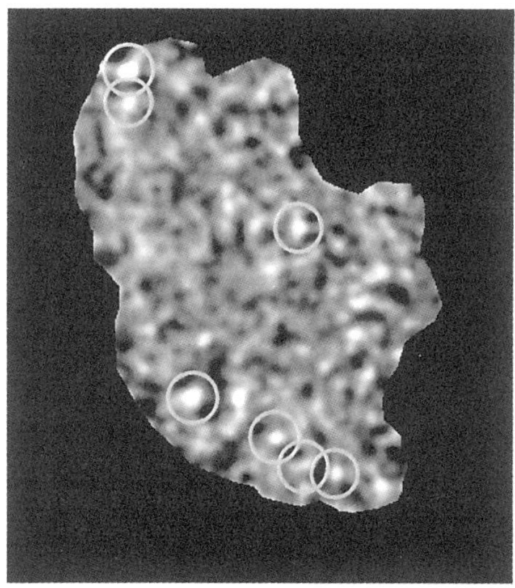

Figure 3. Current SCUBA 850μm map of the ELAISN2 region covering $\simeq$ 100 sq. arcmin. The seven most secure ($\geq 4\sigma$) sources with $S_{850\mu m} > 9$mJy, are marked by circles. The beam signature, with negative side-lobes 45° east-of-north, is visible for the brighter sources.

enshrouded starbursts at high-z can now be carried out. Several such surveys are underway[8,23], but it is clear that reliable source detection, optical/IR identification, and redshift determination is still in its infancy.

One of the main reasons for this somewhat slow progress is the fact that it is only for bright ($S_{850\mu m} > 8$ mJy) SCUBA sources that unconfused positions can be reliably obtained with, for example, follow-up mm interferometry. This is well demonstrated by the effort required to determine an accurate position for the brightest source (HDF850.1; $S_{850\mu m} = 7$ mJy) from our SCUBA survey of the HDF [6]. Indeed, even with sub-arcsec positional accuracy, it can prove hard to distinguish between alternative candidate IDs and, as the test-case of HDF850.1 demonstrates, it is often vital to supplement good astrometric information with SED-based redshift constraints derived from deep radio to far-infrared photometry. In this case it is hard to escape the conclusion that, as for the radio galaxies and their companions, this luminous sub-mm source lies at $z > 3$, in which case its optical counterpart may well be too faint and/or red to be detected in published optical/IR images of the HDF.

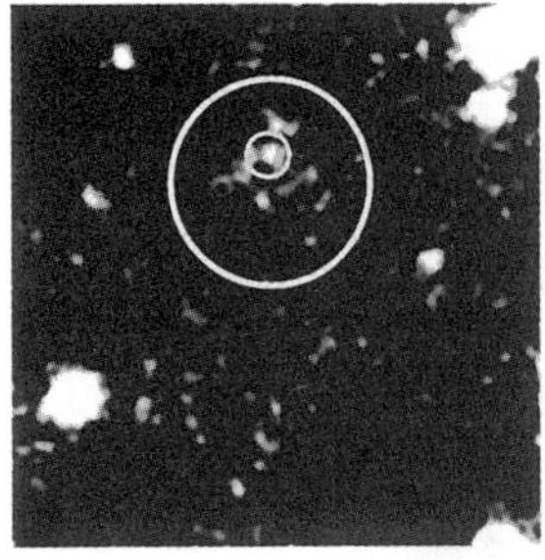
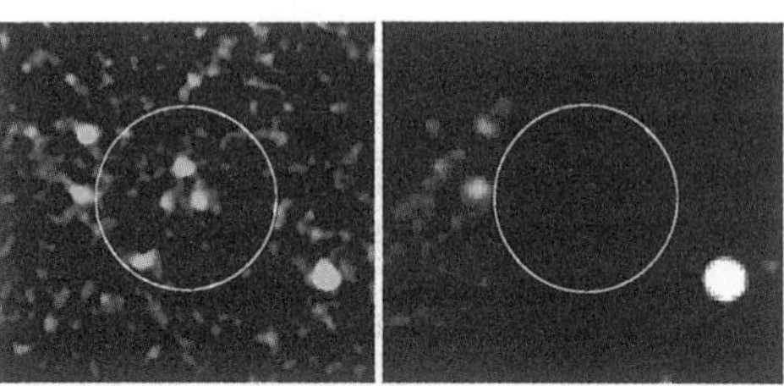

Figure 4. Current information on the two most luminous sub-mm sources detected in the SCUBA 8-mJy survey. The left-hand panel shows a 30×30 arcsec region of a deep (6-hr on UKIRT) K-band image of the field in the vicinity of the $S_{850\mu m} = 13$ mJy source Lockman850.1. The position of the SCUBA source is marked by the large (10-arcsec diameter) circle, while the position of the 1.3mm source detected in follow-up observations with the IRAM PdB interferometer is marked by the small (2-arcsec diameter) circle. With this positional accuracy the SCUBA source can be confidently identified with the brightest of a group of compact peaks with $K \simeq 21$, and $R - K > 5$. The two right-hand panels show the position of the even brighter SCUBA source ELAIS850.1 ($S_{850\mu m} = 20$ mJy) superimposed on a 1-hour K-band UKIRT image (centre panel) and an R-band image reaching $R \simeq 26$. We still await IRAM PdB observations of this object, but again the bright SCUBA source is associated with a group of faint, red knots ($K \simeq 20.5$; $R - K > 5$).

4.1 The 8-mJy 850μm SCUBA Survey

In an effort to assemble a substantial and unbiased sample of sub-mm sources of a luminosity comparable to, or greater than, HDF850.1, we are currently undertaking an $850\mu m$ SCUBA survey of sources brighter than $8\,$mJy in an area of 400 sq.arcmin. In comparison to existing SCUBA surveys this new survey has four key advantages. First, its flux limit is sufficiently bright to allow follow-up with existing instrumentation (e.g. the IRAM PdB interferometer) which should ultimately yield an accurate (~ 1 arcsec) position for every source. Second, there is no a priori reason to expect the sources to be lensed. Third, the flux limit is sufficiently bright that sub-mm confusion is not a problem. Fourth, and scientifically most important, any sources discovered in this survey must be as bright or brighter in the sub-mm than the extreme radio galaxies mentioned above. This survey is therefore optimised for the detection of starbursts with SFR $\simeq 1000\,M_{\odot}yr^{-1}$. Initial results from this survey indicate a cumulative source count of 250 ± 70 degree^{-2} at flux densities $S_{850\mu m} \geq 10\,$mJy, and provide preliminary evidence that these luminous sub-mm sources are strongly clustered (Fig.3).

The crucial next stage is to determine the nature/redshift of the bright

SCUBA sources in this survey. As a first step we have been investigating the two brightest sources detected to date in our 8-mJy survey. These two sources - ELAIS850.1 and Lockman850.1 - have flux densities at $850\mu m$ of 20 and 11 mJy respectively, making them among the very brightest unlensed SCUBA sources discovered from blank-field surveys so far. For Lockman850.1 we have obtained a clear detection at 1.3mm with the IRAM PdB interferometer, yielding its position to sub-arcsec accuracy [16], and similar observations are also planned for ELAIS850.1. We have also obtained deep K-band images of these sources with UFTI on UKIRT. The results of this sub-mm/mm/infrared/optical comparison are shown in Fig.4. In both cases the SCUBA source is associated with a clump of very red objects, with $K \simeq 21$ but $R > 26$, and in the case of Lockman850.1 the IRAM position ties the SCUBA source to the brightest of these clumps. What is so striking about these images is their apparent similarity, and in particular the complexity of the sources at K. We are certainly *not* seeing either an obscured AGN nucleus, or a relaxed elliptical galaxy at intermediate redshift. In fact, these images are very reminiscent of the complex K-band morphologies found for radio galaxies at $z > 3$, reinforcing the connection between radio galaxies and the bright sub-mm population in general [26].

Finally, I note that the one comparably-bright sub-mm source reported from the SCUBA survey of the CFRS [8] has also been identified with an ERO via IRAM interferometry [9], and again has SED constraints indicating $z = 2 \rightarrow 5$. I conclude, therefore, that if attention is confined to the most luminous sub-mm sources, we find rather little 'diversity' in this population. Whether radio galaxies, cluster companions, or blank field sources, all appear to lie at $z > 2.5$, and are associated with (often complex) EROs.

5 Implications

Given their surface density, extreme ($S_{850\mu m} > 10$mJy) sub-mm sources can only contribute 5-10 % of the sub-mm background. However if this population is indeed confined to a relatively narrow redshift band (e.g. $3 < z < 5$), their comoving density is $\simeq 1 - 2 \times 10^{-5} \mathrm{Mpc}^{-3}$, the same as the present-day number density of massive ellipticals of comparable mass to radio galaxies ($\geq 4L^\star$).

For many years there has been growing evidence that the bulk of the stellar populations in radio galaxies formed at high redshift, $z > 3$. I conclude that the available data indicates that radio galaxies are *not* special in this regard. Virtually all known luminous sub-mm sources are confined to this same high-redshift regime, and their number density is sufficient to account for the formation of all massive ($\geq 4L^\star$) ellipticals. These results suggest that CDM-based models need to be tuned to produce very rapid collapse and conversion into stars of the baryonic gas in the most massive haloes [10].

Acknowledgments

I gratefully acknowledge the contributions of my collaborators on the projects covered in this review, especially Elese Archibald, Dave Hughes, Rob Ivison, Steve Rawlings, Omar Almaini, Dieter Lutz, Chris Willott, Raul Jimenez, John Peacock, Suzie Scott and other members of the UK SCUBA Consortium.

References

1. E.N. Archibald *et al*, *MNRAS*, in press, astro-ph/0002083 (2000).
2. C.M. Baugh, S. Cole, C.S. Frenk, *MNRAS* **282**, L27 (1996).
3. N. Benítez *et al*, *ApJ* **515**, L65 (1999).
4. R.G. Bower, T. Kodama, A. Terlevich, *MNRAS* **299**, 1193 (1998).
5. S.C. Chapman *et al*, *MNRAS*, in press, astro-ph/9909092 (2000).
6. D. Downes *et al*, *A&A* **347**, 809 (1999).
7. J.S. Dunlop *et al*, *Nature* **370**, 347 (1994).
8. S.A. Eales *et al*, *AJ*, in press, astro-ph/0009154 (2000).
9. W.K. Gear *et al*, *MNRAS*, in press, astro-ph/0007054 (2000).
10. G.L. Granato *et al*, *MNRAS*, in press, astro-ph/9911304 (2000).
11. D.H. Hughes *et al*, *Nature* **394**, 241 (1998).
12. R.J. Ivison, *MNRAS* **275**, L33 (1995).
13. R.J. Ivison *et al*, *ApJ*, in press, astro-ph/0005234 (2000).
14. R. Jimenez *et al*, *ApJ* **532**, 155 (2000).
15. G. Kauffman and S. Charlot, *MNRAS* **294**, 705 (1998).
16. D. Lutz *et al*, in preparation (2000).
17. J. Magorrian *et al*, *AJ* **115**, 2285 (1998).
18. F. Menanteau *et al*, *MNRAS*, in press, astro-ph/0007114 (2000).
19. L.A. Nolan *et al*, *MNRAS*, in press, astro-ph/0004325 (2000).
20. J.A. Peacock *et al*, *MNRAS*, in press, astro-ph/9912231 (2000).
21. M. Pettini *et al*, *ApJ* **508**, 539 (1998).
22. A. Renzini and A. Cimatti, in press, astro-ph/9910162 (2000).
23. I. Smail *et al*, in press, astro-ph/0008237 (2000).
24. C.C. Steidel *et al*, *ApJ* **519**, 1 (1999).
25. T. Totani and Y. Yoshii, *ApJ* **501**, L177 (1998).
26. W.J.M. van Breugel *et al*, *ApJ* **502**, 614 (1998).
27. S.E. Zepf, *Nature* **390**, 377 (1997).

MULTI-WAVELENGTH OBSERVATIONS OF OBSCURED REGIONS

A. J. BARGER

Institute for Astronomy, University of Hawaii, 2680 Woodlawn Dr., Honolulu, HI 96822, USA

E-mail: barger@ifa.hawaii.edu

With recent *Chandra* observations, at least 75 percent of the $2-10$ keV background is now resolved into discrete sources. We have obtained deep optical, near-infrared, submillimeter, and 20 cm (radio) images, as well as high-quality optical spectra, of a complete sample of 20 hard X-ray sources in a deep *Chandra* observation of the SSA13 field. The thirteen $I < 23.5$ galaxies have redshifts in the range 0.1 to 2.6. Two are quasars, five show AGN signatures, and six are $z < 1.5$ luminous early galaxies whose spectra show no obvious optical AGN signatures. The seven spectroscopically unidentified sources have colors that are consistent with evolved early galaxies at $z = 1.5-3$. Only one hard X-ray source is significantly detected in an ultradeep submillimeter map; its millimetric redshift is in the range $z = 1.2-2.4$. None of the remaining 19 sources are detected in the submillimeter. These results probably reflect the fact that the 850μm flux limits obtainable with SCUBA are quite close to the expected fluxes from obscured AGN. The hard X-ray sources have an average $L_{FIR}/L_{2-10\ \mathrm{keV}} \sim 60$, similar to that of local obscured AGN. The same ratio for a sample of submillimeter selected sources is in excess of 1100, suggesting that their far-infrared light is primarily produced by star formation.

1 Introduction

After more than 35 years of intensive work, the origin of the hard X-ray background (XRB) is still not fully understood. The XRB photon intensity, $P(E)$, can be approximated by a power-law, $P(E) = AE^{-\Gamma}$, where E is the photon energy in keV. The *HEAO1* A-2 experiment[1] found that the XRB spectrum from $3 - 15$ keV is well described by a photon index $\Gamma \simeq 1.4$. At soft X-ray energies ($0.5 - 2$ keV), 70–80 percent[2] of the XRB is resolved into discrete sources by the *ROSAT* satellite. Most of these sources are optically identified[3] as unobscured active galactic nuclei (AGN) with spectra that are too steep to account for the flat XRB spectrum. Thus, an additional population of either absorbed or flat spectrum sources is needed to make up the background at higher energies.

XRB synthesis models, constructed within the framework of AGN unification schemes, were developed to account for the spectral intensity of the XRB and to explain the X-ray source counts in the hard and soft energy bands (e.g., Setti & Woltjer[4]). In the unified scheme, the orientation of a molecular torus surrounding the nucleus determines the classification of a source. The models invoke, along with a population of unobscured type-1 AGN whose emission from the nucleus we see directly, a substantial population of intrinsically obscured AGN whose hydrogen column densities of $N_H \sim 10^{21} - 10^{25}$ cm^{-2} around the nucleus block our line-of-sight.

A significant consequence of the obscured AGN models is that large quantities of dust are necessary to cause the obscuration. The heating of the surrounding gas and dust by the nuclear emission from the AGN and the subsequent re-radiation of this energy into the rest-frame far-infrared (FIR) suggests that the obscured AGN should also contribute to the source counts and backgrounds in the FIR. At high redshifts ($z \gg 1$) the FIR radiation is shifted to the submillimeter.

Due to instrumental limitations, the resolution of the XRB into discrete sources at hard energies had to wait for the arcsecond imaging quality and high-energy sensitivity of *Chandra*. Deep *Chandra* imaging surveys are now detecting sources in the $2 - 10$ keV range that account for 75 to 90 percent[5] of the hard XRB, depending on the XRB normalization. The mean X-ray spectrum of these sources is in good agreement with that of the XRB below 10 keV. Furthermore, because of the excellent $< 1''$ X-ray positional accuracy of *Chandra*, counterparts to the X-ray sources in other wavebands can be securely identified[5,6].

2 SSA13 Hard X-ray Sample

The present study is based on a 100.9 ks X-ray map of the SSA13 field that was observed with the ACIS-S instrument on the *Chandra* satellite in December 1999 and presented in Mushotzky et al.[5] The position RA(2000)= 13^h 12^m 21.40^s, Dec(2000)= $42°$ $41'$ $20.96''$ was placed at the aim point for the ACIS-S array (chip S3). Two energy-dependent images of the back-illuminated S3 chip and the front-illuminated S2 chip were generated in the hard ($2 - 10$ keV) and soft ($0.5 - 2$ keV) bands.

For our hard X-ray sample we selected sources with $2 - 10$ keV fluxes greater than 3.8×10^{-15} erg cm^{-2} s^{-1} that lie within a $4.5'$ radius of the optical axis; there are 20 such sources in the resulting 57 arcmin^2 area.

We successfully obtained redshift identifications for the 13 hard X-ray sources brighter than $I = 23.5$ mag; all are in the redshift range $z = 0 - 3$. The spectra fall into three general categories: (i) 2 quasars, (ii) 5 AGN, and (iii) 6 optically 'normal' galaxies. For the latter category, the AGN are either very weak or undetectable in the optical. The 'normal' galaxies and AGN follow the upper envelope of the star forming field galaxy population.

Thus, the hard X-ray sources predominantly lie in the most optically luminous galaxies.

3 Submillimeter Properties of the Hard X-ray Sample

The new population[7,8,9,10,11,12] of highly obscured, exceptionally luminous sources discovered by SCUBA appear to be distant analogs of the local ultraluminous infrared galaxies[13] (ULIGs). There is an ongoing debate on whether local ULIGs are dominantly powered by star formation or by heavily dust enshrouded AGN, and the same applies to the distant SCUBA sources. Barger et al.[14] carried out a spectroscopic survey of a complete sample of submillimeter sources detected in a survey of massive lensing clusters. Only 4 of the 17 sources could be reliably identified spectroscopically; of these, 3 showed AGN signatures. Thus, the possibility that most SCUBA sources contain AGN remains open. Several authors[15,16] have modelled the X-ray and submillimeter backgrounds; they predict an AGN contribution to the SCUBA surveys at the level of 10 to 20 percent.

The results of recent searches for submillimeter counterparts to *Chandra* X-ray sources have been mixed. In a study of two clusters, Fabian et al.[17] identified three significant $2-7$ keV sources, but only a fourth marginal source was seen in both the X-ray (2.8σ) and submillimeter (2σ) datasets. Likewise, Hornschemeier et al.[18] did not see either of their $2-8$ keV sources in the ultradeep Hubble Deep Field SCUBA map[8]. In contrast, both of the $2-10$ keV sources detected by Bautz et al.[19] in the A370 lensed field are submillimeter sources[7]. These two sources were previously identified spectroscopically as AGN[20,14]. The above mixed results probably reflect the fact that the 850 μm flux limits obtainable with SCUBA are quite close to the expected fluxes from the obscured AGN, as we address below.

The present study has the advantage that wide-area submillimeter[10,21] and extremely deep 20 cm data (Richards et al., in preparation) exist over the entire X-ray field of 57 arcmin^2. An ultradeep 50 hr submillimeter map[9] also exists for one region of the field that contains two hard X-ray sources.

With one exception, the hard X-ray sources are not detected in the submillimeter at the 3σ level. The exception is an optically faint, highly absorbed X-ray source (hereafter, source A) in the ultradeep submillimeter map. The bolometric FIR flux is related to the rest-frame 20 cm flux through the well-established FIR-radio correlation[22] of local starburst galaxies and radio quiet AGN. The spectral energy distributions (SEDs) of submillimeter sources are reasonably well approximated[23,24] by the thermal black-body spectrum of the ULIG Arp 220, which is powered by star formation[25]. We can therefore use

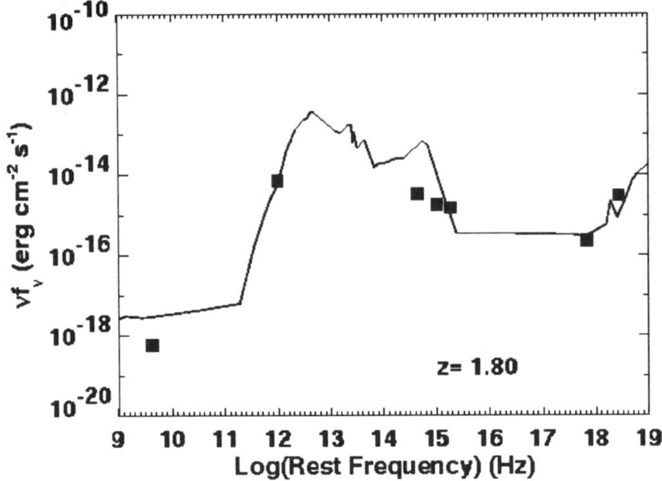

Figure 1. The rest-frame SED of source A (filled squares) at millimetric redshift $z = 1.8$ is shown in comparison with the SED of the heavily dust obscured local AGN NGC 6240 (solid curve).

the submillimeter to radio flux ratio (see Eqs. 2 and 4 of Barger, Cowie, & Richards[24]) to infer an approximate redshift of 1.8 (with estimated redshift range $1.2 - 2.4$) for source A. The millimetric redshift estimation technique is expected to hold for sources dominated by star formation or for radio quiet AGN.

Most of the hard X-ray sources in the sample are consistent with being obscured AGN. In Fig. 1 we compare the rest-frame SED of source A (filled squares) with the SED of the local heavily dust obscured AGN NGC 6240 (solid curve[26]), normalized to approximately match the hard and soft X-ray data of source A.

By considering the ratio of the FIR luminosity, L_{FIR}, to the hard X-ray luminosity, L_{HX}, we can eliminate any redshift or cosmology dependence and make relative comparisons of distant and local systems. The hard X-ray sources with spectroscopic identifications are mostly at modest redshifts where radio observations provide a more sensitive route for obtaining total FIR luminosities than do submillimeter observations[24]. We calculate our FIR luminosities by assuming that the sources are well described by the FIR-radio correlation given in Sanders & Mirabel[13]. Three of the spectroscopically identified hard X-ray sources are not detected above the 15 μJy (3σ) limit in the 20 cm data, so we calculate their luminosities as upper limits based on the 3σ radio limit.

To allow for the average effects of opacity, we calculate our hard X-ray luminosities by normalizing the flux at 4 keV for an assumed $\Gamma = 2$ unabsorbed spectrum to the flux at 4 keV calculated over the $2 - 10$ keV energy range for a spectrum with $\Gamma = 1.2$. ($\Gamma = 1.2$ is the counts-weighted mean photon index for our hard band sample[5].)

Figure 2a shows the L_{FIR}/L_{HX} ratio versus L_{HX} for our data sample. Following Fabian et al.[17], we also show on the figure the data values for the local ultraluminous infrared galaxies[27,28] Arp 220 and NGC 6240, the $z = 0.4$ radio quiet quasar[29,30,31] PG 1543+489, and the radio loud quasar[32,33] 3C 273. NGC 6240 hosts a powerful AGN and is highly absorbed with an inferred column density of $N_H \sim 2 \times 10^{24}$ cm^{-2}. On the figure we have connected with a straight line the values of L_{HX} calculated from the observed flux[28] and from the inferred intrinsic flux[34] to illustrate the large effects that absorption can make for highly obscured AGN at low redshift.

We find that the values of L_{FIR}/L_{HX} for all the sources are comparable to that of NGC 6240. Also shown in Fig. 2a (open diamonds) are the luminous $z > 4$ radio quiet quasars observed in the submillimeter[35] for which there are also X-ray detections[36]. These quasars are in the region of 3C 273 in Fig. 2a. The two gravitationally-lensed sources[7,19] in the field of A370 (open triangles) have higher L_{FIR}/L_{HX} ratios than the typical X-ray source. However, none of the hard X-ray selected sources remotely approaches the high $L_{FIR}/L_{HX} = 3.4 \times 10^4$ value of Arp 220.

Figure 2b shows L_{FIR}/L_{HX} versus L_{HX} for submillimeter sources with millimetric redshifts[24,21] and X-ray[18,21] detections or limits. If most of the submillimeter sources are forming stars at a high rate like Arp 220, then much deeper hard X-ray observations would be required to detect them.

4 Hard X-ray Properties of a Submillimeter Selected Sample

What are the properties of a submillimeter selected sample? There are twelve 3σ source detections at 850 μm within a 4.5' radius of the optical axis. The fluxes range from just over 2.3 mJy to 11.5 mJy. The positional accuracies for

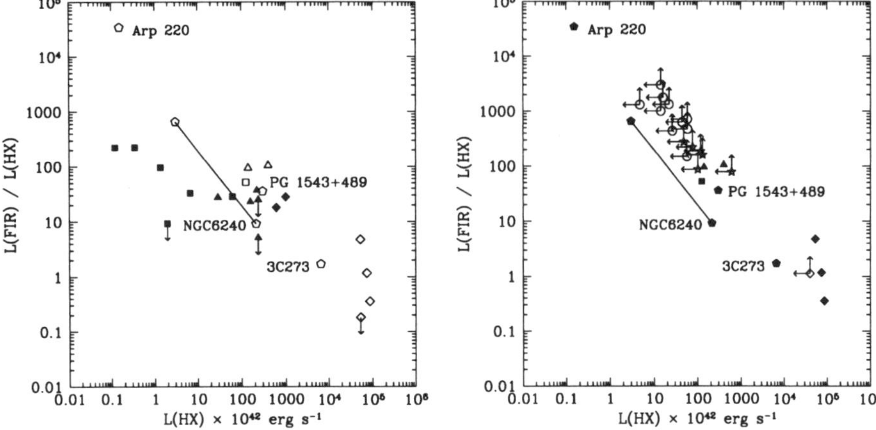

Figure 2. L_{FIR}/L_{HX} versus L_{HX} for the hard X-ray sample with redshifts. Quasars are denoted by filled diamonds, AGN by filled triangles, 'normal' galaxies by filled squares, and source A by an open square. Also shown in the figure are the data values for the local ULIGs Arp 220 and NGC 6240, the radio quiet quasar PG 1543+489, and the brightest quasar in the sky, 3C 273. A straight line has been drawn between the absorption-corrected (higher X-ray luminosity) and observed NGC 6240 ratios to illustrate the large effects absorption can make at low redshift. The luminous $z > 4$ radio quiet quasars detected in both the submillimeter and X-ray are shown by the open diamonds. The two significant X-ray and submillimeter sources in A370 are shown by the open triangles. (b) L_{FIR}/L_{HX} versus L_{HX} for $> 3\sigma$ submillimeter sources with millimetric redshifts and hard X-ray detections or limits. Open circles denote the sources in the Hubble Deep Field and Flanking Fields, and open stars denote the sources in the SSA13 field. Symbols for the comparison sources and for the sources significantly detected in the X-ray from (a) have been filled in for clarity.

the submillimeter sources are relatively poor because of the large beam size. In order to allow for the positional uncertainty, we determined the $2 - 10$ keV fluxes for each of the submillimeter sources in $10''$ diameter apertures. The only submillimeter source that is also a strong hard X-ray emitter is source A. The submillimeter source with the second strongest hard X-ray flux is not in the hard X-ray sample but is a known soft X-ray emitter (source 28 in Table 1 of Mushotzky et al.[5]).

The ratio of the total hard X-ray flux in the sample to the total submillimeter flux is $1.4 \pm 0.5 \times 10^{-16}$ erg cm^{-2} s^{-1} mJy^{-1}. We may use this ratio to estimate the fraction of the hard XRB that arises from submillimeter sources in this flux range. The EBL of the $2 - 10$ mJy source population[10] is 9.3×10^3

mJy deg^{-2}, which would contribute 1.3×10^{-12} erg cm^{-2} s^{-1} deg^{-2} in the $2-10$ keV band or 6 percent of the hard XRB, if we adopt the Vecchi et al.[37] value of 2.3×10^{-11} erg cm^{-2} s^{-1} deg^{-2}.

However, nearly all of the X-ray signal from the submillimeter sample is coming from source A. If this single source is removed, the total hard X-ray to total submillimeter flux ratio drops to $0.6 \pm 0.5 \times 10^{-16}$ erg cm^{-2} s^{-1} mJy^{-1}. If we place the submillimeter sources at $z = 2$ (consistent with the $z = 1-3$ spectroscopic[14] and millimetric[24] redshift range for submillimeter sources), then the 1σ lower limit on L_{FIR}/L_{HX} would be approximately 1100. This lower limit is above the obscured AGN values and approaching that of Arp 220. It therefore appears that most of the submillimeter sources, at least above 2 mJy, are star formers with a small admixture of obscured AGN.

Acknowledgments

I thank James Lowenthal and the other organizers for inviting me to this stimulating conference and for providing support. I thank L. Cowie, R. Mushotzky, E. Richards, and K. Arnaud for collaborations on the work reported here. This work was supported by NASA through Hubble Fellowship grant HF-01117.01-A awarded by the Space Telescope Science Institute, which is operated by the Association of Universities for Research in Astronomy, Inc., for NASA under contract NAS 5-26555.

References

1. F. Marshall *et al*, *ApJ* **235**, 4 (1980).
2. G. Hasinger *et al*, *A&A* **329**, 482 (1998).
3. M. Schmidt *et al*, *A&A* **329**, 495 (1998).
4. G. Setti and L. Woltjer, *A&A* **224**, L21 (1989).
5. R.F. Mushotzky, L.L. Cowie, A.J. Barger, and K.A. Arnaud, *Nature* **404**, 459 (2000).
6. W.N. Brandt *et al*, *AJ* **119**, 2349 (2000).
7. I. Smail, R.J. Ivison, and A.W. Blain, *ApJ* **490**, L5 (1997).
8. D.H. Hughes *et al*, *Nature* **394**, 241 (1998).
9. A.J. Barger *et al*, *Nature* **394**, 248 (1998).
10. A.J. Barger, L.L. Cowie, and D.B. Sanders, *ApJ* **518**, L5 (1999).
11. S. Eales *et al*, *ApJ* **515**, 518 (1999).
12. S.J. Lilly *et al*, *ApJ* **518**, 641 (1999).
13. D.B. Sanders and I.F. Mirabel, *ARA&A* **34**, 749 (1996).
14. A.J. Barger *et al*, *AJ* **117**, 2656 (1999).

15. O. Almaini, A. Lawrence, and B.J. Boyle, *MNRAS* **305**, 59 (1999).

16. K. Gunn and T. Shanks, *MNRAS*, in press (2000).

17. A.C. Fabian *et al*, *MNRAS* **315**, L8 (2000).

18. A.E. Hornschemeier *et al*, *ApJ*, in press (2000).

19. M.W. Bautz *et al*, *ApJL*, submitted (2000).

20. R. Ivison *et al*, *MNRAS* **298**, 583 (1998).

21. A.J. Barger, L.L. Cowie, R.F. Mushotzky, and E.A. Richards, *AJ*, submitted (2000)

22. J.J. Condon, *ARA&A* **30**, 575 (1992).

23. C.L. Carilli and M.S. Yun, *ApJ* **530**, 618 (2000).

24. A.J. Barger, L.L. Cowie, and E.A. Richards, *AJ* **119**, 2092 (2000).

25. D. Downes and P.M. Solomon, *ApJ* **507**, 615 (1998).

26. G. Hasinger in *ISO Surveys of a Dusty Universe*, eds. D. Lemke, M. Stickel, K. Wilke (Springer), in press (2000).

27. U. Klaas, M. Haas, I. Heinrichsen, B. Schulz, *A&A* **325**, L21 (1997).

28. K. Iwasawa, *MNRAS* **302**, 96 (1999).

29. M. Polletta *et al*, *A&A*, in press (2000).

30. A. Laor *et al*, *ApJ* **477**, 93 (1997).

31. S. Vaughan, J. Reeves, R. Warwick, R. Edelson, *MNRAS* **309**, 113 (1999).

32. D.-C. Kim and D.B. Sanders, *ApJS* **119**, 41 (1998).

33. M. Türler, T.J.-L. Courvoisier, S. Paltani, *A&A* **349**, 45 (1999).

34. P. Vignati *et al*, *MNRAS* **308**, L6 (1999).

35. R. McMahon *et al*, *MNRAS* **309**, L1 (1999).

36. S. Kaspi, W.N. Brandt, and D.P. Schneider, *AJ* **119**, 2376 (2000).

37. A. Vecchi *et al*, *A&A* **349**, L73 (1999).

WIDE FIELD IMAGING AT 250 GHZ

C.L. CARILLI, F. OWEN, M. YUN

NRAO

F. BERTOLDI, A. BERTARINI, K.M. MENTEN, E. KREYSA, R. ZYLKA

MPIfR

We summarize results from sensitive, wide-field imaging using the Max-Planck Bolometer Array at the IRAM 30m telescope, including source counts, clustering, and redshift distribution.

The Max-Planck mm Bolometer Array (MAMBO) is a 37 element, 250 GHz bolometer array operating at 300 mK. We have observed extensively with MAMBO at the IRAM 30m telescope over the last two winters. In this paper we summarize the preliminary results from the wide field imaging programs with r.m.s. sensitivities of ~ 0.5 mJy over areas ~ 200 sq. arcmin.

The three fields observed thus far are the NTT Deep field, the Lockman Hole, and the $z = 0.25$ cluster Abell 2125. In parallel with the wide field imaging at 250 GHz, we have observed, or will be observing, these fields with the VLA at 1.4 GHz to rms $= 7$ μJy, and we are obtaining sensitive optical and near-IR wide field images with various optical telescopes.

Figure 1 shows the MAMBO image of the Abell 2125 field. We detect 36 sources with 250 GHz flux densities $S_{250} \geq 2$ mJy in regions with rms noise ≤ 0.6 mJy. Follow-up observations of a few of the MAMBO sources in the Abell 2125 field[2] suggests they are the same population as the SCUBA sources, *i.e.* very faint in the optical and near-IR, with $K \geq 20.5$.

Figure 2a shows the cumulative source counts based on two of the three MAMBO fields, along with source counts determined from various SCUBA surveys. We relate 250 GHz flux densities to 350 GHz flux densities using a scaling factor of 2.25. This factor is applicable to a typical starburst galaxy at $z \approx 2.5$. We have included faint source counts in the regions within a $1'$ radius of the cluster center assuming a mean gravitational magnification factor of 2.5. The MAMBO and SCUBA counts agree well at intermediate flux densities ($S_{350} = 2$ to 8 mJy). The very wide fields imaged by MAMBO allow us to set the best constraints to date at high flux densities, and we find that there is steepening in the distribution at $S_{350} = 10$ mJy, consistent with an exponential cut-off in the starburst galaxy population at about 10^{13} L$_\odot$. All of the data can be reasonably fit by a Schechter-type luminosity function, with a power-law index of -2 and an exponential cut-off at 10 mJy.

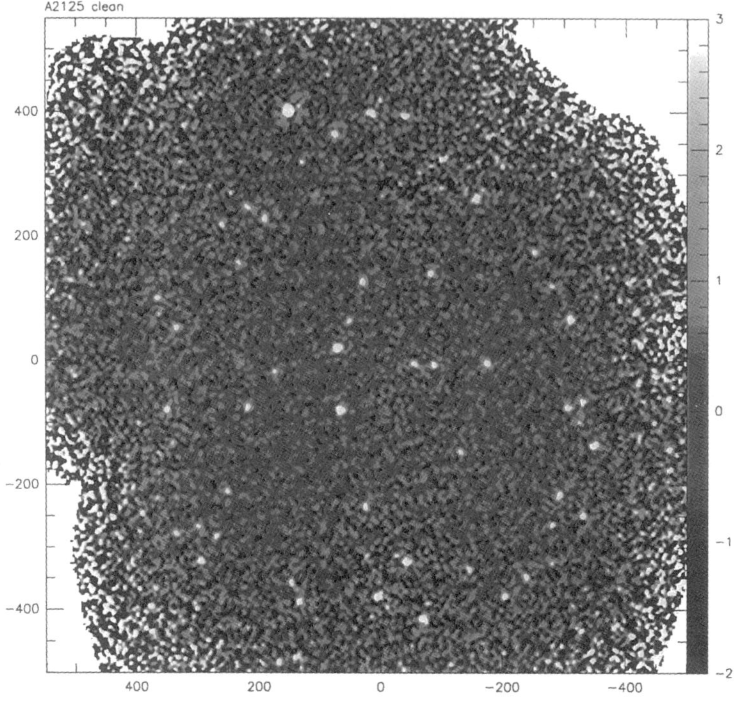

Figure 1. The MAMBO image at 250 GHz of the Abell 2125. The angular scale is in arcseconds, and the greyscale range is in mJy, and the rms noise = 0.5 mJy.

Figure 3a shows the redshift distribution for the Abell 2125 field sources that are detected at both 250 GHz and 1.4 GHz. We selected this field for a redshift analysis since it has the deepest radio image of the three MAMBO fields, with an rms = 7 μJy. The redshifts were determined using the 1.4 GHz-to-250 GHz flux density ratio applicable to star forming galaxies[1]. We find that 18 of the 36 sources with $S_{250} \geq 2$ mJy are detected at $S_{1.4} \geq 22\mu$Jy. The median redshift is 2.5, with most of the sources between $2 \leq z \leq 4$. The high radio detection rate gives us confidence in the reliability of the MAMBO selected sources. We expect only about 1 source to have been detected at random at 1.4 GHz within 3$''$ of any of the 36 MAMBO source positions. The high detection rate also implies that the sensitivities of the mm and cm surveys

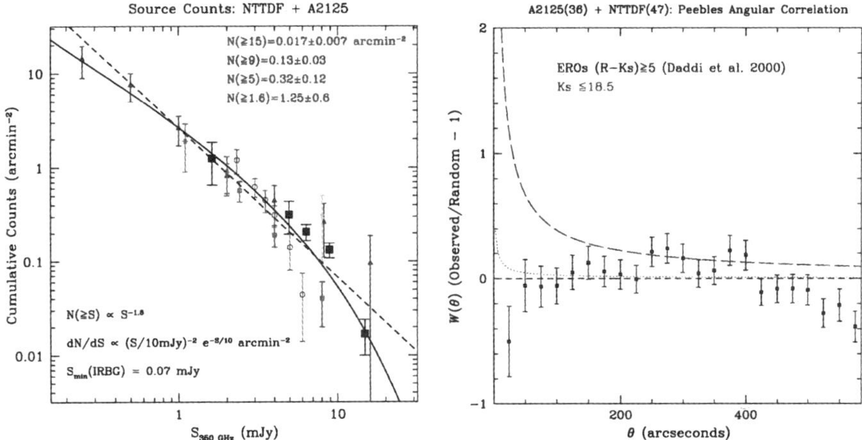

Figure 2. The figure on the left shows the source counts from the MAMBO fields as large solid squares, plus counts from various SCUBA surveys (this volume). The dashed curve is a power-law of index −1.6. The solid curve is an (integrated) Schechter function with parameters as given in the text and on the plot. The figure on the right shows the two point correlation function for 250 GHz sources in the NTT deep field and the Abell 2125 field. The dashed curve is for EROs and the dotted curve is for galaxies with $K_s \leq 18.5$.

are well-matched, and that there is not a dominant 'hidden' population of sources, either at very high redshift, or at low redshift but with radio flux densities below those expected for starburst galaxies based on the radio-to-far IR correlation. Given these results, it is likely that the upcoming radio imaging programs with the VLA pushing down to an rms $\approx 3\mu$Jy at 1.4 GHz will result in a close-to-complete radio detection rate for fields such as this.

Figure 3a also shows the redshift distribution for SCUBA sources at 350 GHz with radio detections. This shows a lower median redshift of 1.9. A systematic offset in redshift between sources selected at 250 GHz relative to 350 GHz is expected qualitatively, since the spectral energy distributions go 'over-the-top' of the IR peak at lower redshift at 350 GHz. Quantitatively however, one would expect this segregation to occur at higher redshift, $z \approx 6$, hence we feel that the segregation in redshift in Figure 3a is more likely due to the deeper radio survey for Abell 2125.

Figure 3b shows the total redshift distribution for sources selected at either 250 or 350 GHz, including sources with spectroscopic redshifts, radio detections, and radio lower limits. The distribution is broad, but most of the sources appear to be in the range $1.5 < z < 4$, with a median of 2.3 for

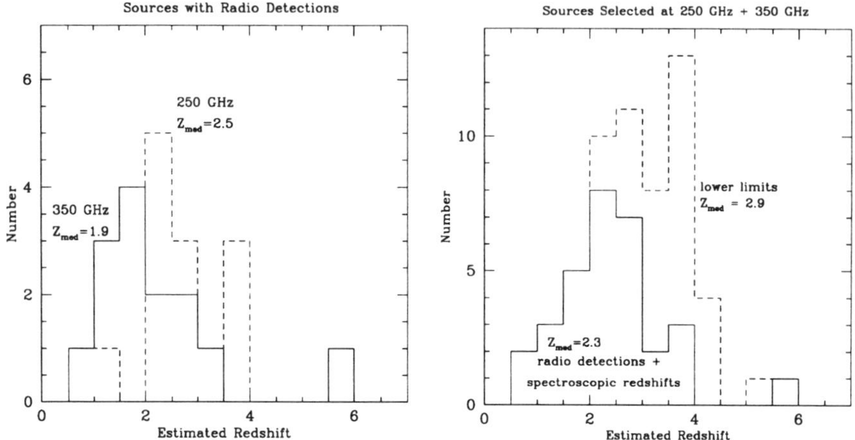

Figure 3. The left figure shows the redshift distribution for 250 GHz and 350 GHz selected sources separately, based on the cm-to-mm flux density ratio for sources with radio detections at 1.4 GHz. The right figure shows the distribution for all the sources, including redshift lower limits based on upper-limits to the radio flux density.

sources with radio detections and/or spectroscopic redshifts, and 2.9 if radio lower limits are included. The lower limits leave open the possibility of a substantial, although not majority, population of high-z sources, with $z > 3$.

Figure 2b shows the clustering properties of the sources in the Abell 2125 plus the NTT deep field. Plotted is the Peebles two-point correlation function, determined using a random distribution of a large number of sources with the same spatial sampling as the MAMBO fields. We also plot the clustering properties of Extremely Red Objects (EROs)[3]. The results are noisy, due to the small number of sources, but the mm-selected sources appear to be less clustered than the EROs. This is not surprising, since the EROs are mostly intermediate-redshift elliptical galaxies with very strong clustering properties. Also, given the very flat 'redshift selection function' for mm sources from $z = 0.5$ to 7, any clustering will be highly diluted by the large volume sampled.

References

1. Carilli, C.L., Yun, M.S., *ApJ* **530**, 618 (2000).
2. Bertoldi, F. *et al*, *A&A* **360**, 92 (2000).
3. Daddi, E. *et al*, *A&A*, in press (2000).

2. Templates at Low Redshift

MOLECULAR GAS IN GALAXIES IN THE LOCAL UNIVERSE

JUDITH S. YOUNG

Dept. of Astronomy and FCRAO, University of Massachusetts,
Amherst, MA 01003 USA
E-mail: young@astro.umass.edu

The observations made as a part of the FCRAO Extragalactic CO Survey are used to examine the molecular gas distributions, the global H_2 masses, the star formation efficiency, the H_2/HI ratio, and the gas to dust ratio, within and among 300 galaxies in the local Universe as a function of morphology, luminosity, and environment. Significant global trends in the gas-to-dynamical mass ratio, the molecular gas-to-warm dust mass ratio, and the H_2/HI ratio are found as a function of morphology for spiral galaxies along the Hubble sequence. These results are discussed and used to provide insight into the processes of star formation and galaxy evolution, and to place constraints on current models of the evolution of galaxies in the local Universe.

1 Introduction

Studies of molecular clouds are essential to our understanding of the morphology and evolution of galaxies, since it is within the dense molecular clouds that new generations of stars form, and it is the young high mass stars that produce a major part of the galactic luminosity. Furthermore, the observations of molecular clouds in galaxies in the local Universe ($z < 0.03$) provide the basis for interpreting the molecular gas and dust properties of higher redshift systems.

The molecular cloud observations used in this paper are from the FCRAO Extragalactic CO Survey (Young et al. 1995), in which the CO J=1$\rightarrow$0 emission was observed at 1412 positions in 300 galaxies with 45" resolution. These observations make up the most extensive, uniform database on the radial distributions of molecular gas and the global molecular gas content of spiral galaxies in the local Universe. With this database, we can statistically address the global H_2 content of galaxies in relation to other components in the disk – specifically the atomic gas content, the warm dust mass, and the rate of high mass star formation – as a function of morphological type, luminosity, and environment.

For the majority of galaxies in the CO Survey, the 45" half-power beam width of the 14 m telescope corresponds to a linear resolution of 1-5 kpc. Therefore, the molecular gas surface densities we have measured correspond to regions within galaxies that are larger than the sizes of giant molecular clouds, and represent large-scale average surface densities. The distances to the galaxies in our sample were derived assuming a Hubble constant of 50 km sec^{-1} Mpc^{-1}. The H$_2$ masses used here were derived from major axis CO observations, assuming a constant CO $\to$ H$_2$ proportionality of 2×10^{20} cm^{-2}/[K(T$_R$) km s^{-1}]. The global H$_2$ masses derived for luminous spiral galaxies have been shown to be statistically accurate to $\sim$40% (Devereux and Young 1990), which is comparable to the measurement uncertainty in the global CO fluxes (Young et al. 1995). The absolute value of the CO $\to$ H$_2$ proportionality for the Milky Way has been shown to hold also for the disks of such diverse spiral galaxies as M31 (Sb) and M33 (Scd) from observations that resolve individual molecular clouds (see Young and Scoville 1991, Figure 1). For galaxies of low mass and metallicity, especially dwarf irregular galaxies, the Galactic value of the CO $\to$ H$_2$ conversion is likely to underestimate H$_2$ masses (Maloney and Black 1988; Wilson 1995).

2 The Global Star Formation Rate and Efficiency in Galaxies

2.1 The Origin of the Far-IR Luminosity in Spiral Galaxies

The origin of the far-IR luminosity within spiral galaxies is a subject of great interest. Of primary importance is the extent to which the far-IR flux densities measured for thousands of galaxies by IRAS can be used to indicate high mass star formation rates. Two primary sources have been proposed to explain the heating of the dust which produces the far-IR luminosity in spiral galaxies – young stars and HII regions (cf. Becklin, Fomalont, and Neugebauer 1973; Telesco and Harper 1980; Young et al. 1984, 1986b, 1996; Helou, Soifer, and Rowan-Robinson 1985; Sanders and Mirabel 1985; Stark et al. 1986; Solomon and Sage 1988; Devereux and Young 1990a, 1993) and non-ionizing stars (Lonsdale and Helou 1987; Bothun et al. 1989), with some authors suggesting contributions from both sources (Habing et al. 1984; Rice et al. 1990; Sauvage and Thuan 1992; Smith et al. 1994).

There are several arguments that suggest that the global IR emission from spiral galaxies provides a measure of the high mass star formation rate. First, the dust temperatures deduced for galaxies from the ratio of 60 to 100 μm flux densities are typically 30-40 K, which is similar to the temperature of dust in Galactic star-forming regions (see Wynn-Williams and Becklin 1974 and references therein; Scoville and Good 1989), and considerably greater than the 15-20 K temperatures expected for dust heated by the ambient interstellar radiation field (Jura 1982; Mezger et al. 1982; Draine and Lee 1984; Cox, Krugel, and Mezger 1986). Second, a comparison of the global Hα and far-IR luminosities for 200 spiral galaxies demonstrates that the luminosity measured in the far-infrared is in quantitative agreement with that expected from the O and B stars that are required to ionize the hydrogen gas (Devereux and Young 1990a). Furthermore, we have demonstrated for 120 galaxies that both the Hα and far-IR luminosities trace the same trends in the SFE (Young et al. 1996).

2.2 *The Global Star Formation Efficiency in Galaxies*

One of the most definitive ways to describe the current epoch star formation in a galaxy is through the comparison of the current high mass star formation rate (SFR) with the mass of gas available to form stars. This yield of young stars per unit mass of molecular gas is what we call the star formation efficiency (SFE). Recent determinations of the SFE in hundreds of galaxies have shown that similar results are obtained on average, independent of whether one uses star formation rates traced by Hα emission [L(Hα)] or by the far-IR emission [L(IR)] observed by IRAS (Young et al. 1996). Through these studies, we have addressed the trends in the SFE with morphology along the Hubble sequence, the effects of environment, and trends with galaxy size, as discussed below.

For galaxies of differing morphology, examination of the mean global SFE for spiral galaxies along the Hubble sequence indicates that there is no variation in the mean SFE (based on both $L(IR)/M(H_2)$ and $L(Hα)/M(H_2)$ to trace the star formation efficiency) for morphological types Sa through Sc (Rengarajan and Verma 1986; Young and Scoville 1991; Devereux and Young 1991; Young et al. 1996). For spiral galaxies of types Sa, Sb and Sc, the mean star formation efficiency within each type has a value of $L(IR)/M(H_2)$ of 4 ± 1 $L_\odot/M_\odot$. Thus, even though the galaxies have a diverse morphology and bulge size, the observed constancy of the mean SFE with spiral type indicates a remarkable similarity in the global star formation process in the disks of spiral galaxies.

The most significant enhancement in the global star formation efficiency is found among merging and strongly interacting galaxies, where the SFE is observed to be elevated by a factor of $\sim$5-20 relative to isolated galaxies (Sanders and Mirabel 1985; Young et al. 1986a; Solomon and Sage 1988; Tinney et al. 1990). Furthermore, not only is the SFE elevated, but the temperature of the dust detected by IRAS is elevated in these systems as well (Young et al. 1986b). The increase of the SFE with dust temperature goes roughly as temperature to the power 4.5 ± 1, which is what one expects for thermal emission from dust. The dependence of the SFE on the dust temperature is consistent with a scenario in which the dust in molecular clouds is heated by young stars, and as the luminosity in young stars per unit mass of molecular gas (i.e. the SFE) increases, the resulting increase in the energy density of the radiation field heats the dust to higher temperatures. It is interesting that the ratio of far-infrared luminosity to *atomic* gas mass does not increase with dust temperature (as shown in Young et al. 1989), supporting the suggestion that it is primarily the dust in *molecular* clouds that is radiating at the 30-40 K dust temperatures indicated by the IRAS data.

A wide spread in the SFE is observed at all dust temperatures; among the merging and closely interacting galaxies, the SFE ranges from 4 to 60 $L_\odot/M_\odot$. This scatter is considerably larger than the overall measurement uncertainties ($\sim$40%), and is possibly related to the age of a merger (Young et al. 1986b). Joseph and Wright (1985) have suggested that the age of a merger can be estimated qualitatively based on the presence of two distinct disks in the younger systems (such as NGC 4038/39 and NGC 520), and that as a merger ages, the nuclei are found closer together, until the merger remnant resembles a single galaxy (such as NGC 3310). Since many of the galaxies in our sample are also in the sample of Joseph and Wright (1985), we have used their estimates of merger age in relation to the values of $L(IR)/M(H_2)$. We find that the lowest SFEs are found in the youngest systems, and the highest SFEs in the oldest systems, consistent with the star formation efficiency increasing with merger age as the H_2 is consumed and young stars are formed.

The enhanced SFE in the interacting/merging galaxies relative to isolated galaxies could reflect the initiation of a new mechanism for the formation of high mass stars. Alternatively, the physical process which causes stars to form in interacting galaxies may be no different from that in isolated galaxies, but the interaction may increase its effectiveness. One such mechanism is that of cloud-cloud collisions, which has been shown to become important during galaxy-galaxy interactions. Noguchi and Ishibashi (1986) have made numerical simulations of galaxy-galaxy interactions including gas clouds as well as stars. They find that the cloud-cloud collision rate increases as bridges and tails develop during violent encounters, reaching a maximum value of $\sim$8 times the pre-encounter value $\sim 3\times10^8$ yr after closest approach of the perturber. If cloud-cloud collisions are responsible for high mass star formation in galaxies, the enhanced rates of cloud-cloud collisions should result in increased SFEs.

Even among the isolated galaxies in our sample, which are primarily types Sbc-Scd, there is a spread of a factor of 10 in the observed SFE for galaxies of a given dust temperature. The origin of some of the spread in the star formation efficiencies among galaxies of all types appears to be galaxy size (Young 1999). For isolated galaxies, Virgo galaxies, pairs, and field spirals observed in the FCRAO Extragalactic CO Survey, there is a trend of decreasing mean SFE with increasing galaxy size. For 18 galaxies smaller than 16 kpc in diameter, the mean SFE is 9 ± 1 $L_\odot/M_\odot$, while for 25 galaxies larger than 50 kpc in diameter, the mean SFE is 3 ± 0.5 $L_\odot/M_\odot$. This same trend is found among subsets of galaxies of each Hubble type (Young 1999). One possible explanation of this trend is that it may result from a greater shear in the disks of large galaxies, where flat rotation curves are found, in contrast to the disks of smaller galaxies, where rotation curves are rising over most of the disk (Rubin et al. 1985). This shear in the larger galaxies would be expected to increase the turbulent energy in molecular clouds, and possibly reduce the SFE.

3   HI and H_2 Masses in Galaxies

Prior to the formation of stars from molecular clouds, the molecular clouds themselves must form from the reservoir of atomic gas. While determinations of the atomic gas content of galaxies have been underway for 40 years, the relative youth of the field of extragalactic molecular studies has meant that knowledge of the relative amounts of molecular and atomic gas in galaxies is limited to the galaxies observed in CO. For 178 galaxies of types Sa-Sc, the dominant neutral gas phase changes along the Hubble sequence, with the mean ratio of molecular to atomic gas ranging from 2 ± 1 among Sa galaxies to 0.4 ± 0.2 for the Sc galaxies (Young and Knezek 1989; Sage 1993). This is consistent with the results found for E, S0, and S0a galaxies when upper limits are included (Thronson et al. 1989; Lees et al. 1991). A somewhat lower molecular to atomic gas ratio for the early type spirals is reported by Casoli et al. (1998). Clearly, different samples of galaxies may be characterized by different ratios of molecular to atomic gas.

For the galaxies in the FCRAO Extragalactic CO Survey, the mean ratio of the total ISM gas mass to optical area is 8 $M_\odot/pc^2$ for the Sa galaxies, and 16 $M_\odot/pc^2$ for the Sc galaxies. Additionally, the molecular gas surface densities are roughly constant from type Sa to Sc, at $\sim$5-6 $M_\odot/pc^2$. Thus, the changing H_2/HI ratio and the increasing gas surface density from type Sa to Sc is caused by an increase in the average HI surface densities from early to late-type spiral galaxies. Relative to the dynamical masses indicated by the rotation velicities in Sa-Scd galaxies, the gas mass fraction (H_2+HI) constitutes 4% of a galaxy's mass for Sa-Sab galaxies, increasing smoothly along the Hubble sequence to to 25% for the Scd galaxies.

4   The Gas to Dust Ratio in Galaxies

One important tracer of the evolution of galaxies is the gas-to-dust ratio. Clearly, this quantity will evolve with time, since the gas mass decreases over time and the dust abundance should increase. Based on a comparison of the global H_2 masses with the warm dust masses that are required to produce the emission measured by IRAS, studies of over 100 spiral galaxies have consistently found molecular gas-to-warm dust mass ratios of $\sim$600 (Young et al.

1986b, 1989; Stark et al. 1986; Devereux and Young 1990b; Sanders et al. 1991). This is in contrast to the value of the gas-to-dust ratio of $\sim$100-150 that is widely used for the Galaxy (Hildebrand 1983; Draine and Lee 1984). The IRAS data lead to similarly high gas-to-warm dust ratios for individual molecular clouds in the Galaxy as well (Heyer et al. 1989). It is noteworthy that inclusion of the atomic gas mass with the molecular gas mass will only accentuate the discrepancy, particularly for spiral galaxies in which the atomic gas mass is the dominant component in the ISM. One plausible explanation for the apparently high gas-to-dust ratios in external galaxies when compared with the Milky Way is that IRAS is not sensitive to the the bulk of the dust mass, which is radiating beyond 100 μm. For this reason, we refer to the dust detected by IRAS as "warm dust."

4.1 The Remarkable Molecular Gas-to-Warm Dust Correlation

Based on the analysis of of the molecular gas-to-warm dust mass ratio in 124 spiral galaxies, Young et al. (1989) found that the molecular gas masses were highly correlated with the warm dust masses derived from the IRAS data (correlation coefficient = 0.97) for galaxies covering a wide range of morphology and luminosity. This correlation is tighter even than the well-known radio-infrared correlation. For our sample, the molecular gas-to-warm dust mass ratio was found to be 570±50, with a remarkably small scatter. No differences were found in the molecular gas-to-warm dust mass ratio as a function of morphology or luminosity. In sharp contrast, the ratio of atomic gas-to-warm dust was found to exhibit a considerably greater amount of scatter (correlation coefficient = 0.79).

For the subset of galaxies for which maps of the HI distribution are available in the literature, it is possible to determine the total gas-to-warm dust ratio in the inner disk, i.e. the area where most of the star formation and dust heating occurs. For 58 spiral galaxies examined by Devereux and Young (1990b), the ratio of the mass of inner disk atomic plus molecular gas to the mass of warm dust has a mean value of 1080±70. If the total gas-to-total dust ratio in external galaxies is the same as the value of $\sim$100-150 determined for the Galaxy, then the high value of 1080 derived from the IRAS data indicates that only 10-20% of the total dust mass is warm enough to radiate in the IRAS bands. Conversely, 80-90% of the dust mass must be radiating at wavelengths beyond 100 μm and is therefore colder than 30 K.

The observed high value of the gas-to-dust ratio indicates that it is clearly erroneous to derive gas masses for galaxies with only an observed infrared flux density and an assumed gas-to-dust ratio, especially for galaxies in the early Universe. The reason for this is twofold. First, one has no a-priori knowledge of the actual gas-to-dust ratio in galaxies at early epochs in the overall star formation history. And second, even if one knew the intrinsic gas-to-dust ratio, deriving a gas mass also requires knowledge of the fraction of the total dust mass which is radiating at the observed wavelength. For nearby galaxies observed by IRAS, the above technique of deriving gas masses from dust masses led to values that were a factor of 10 too high, reflecting the fact that the observations are sensitive to only 10% of the total dust mass.

For a handful of spiral galaxies, including M51 and NGC 6946, maps of the far-IR emission at wavelengths between 160 and 360 μm are available (Smith 1982; Stark et al. 1989; Hunter et al. 1989; Haas et al. 1998; Krugel et al. 1998; Alton et al. 1998), allowing us to examine the possibility that spiral galaxies contain large quantities of cold dust. Although on the one hand, these data provide little direct evidence for the presence of large quantities of cold dust in galaxies, the present observations are consistent with the possibility that most of the dust mass in these galaxies is colder than 15K and radiating at wavelengths longward of 100 μm (Devereux and Young 1992; 1993). In all of these galaxies, due to the strong temperature dependence of thermal emission (T^5 for a λ^{-1} emissivity law), even with 90% of the dust at a temperature of 15 K, the small fraction of the dust at 30 K will radiate 75% of the dust luminosity. Thus, for galaxies in the early Universe, knowledge of the dust temperature and the intrinsic gas-to-dust ratio will be essential for understanding objects and unraveling their gas consumption history.

References

1. Alton, P., *et al.* (1998), *Astr.Ap.*, **335**, 807.
2. Becklin, E., *et al.* (1973), *Ap.J.(Letters)*, **181**, 27.
3. Bothun, G., Lonsdale, C., and Rice, W. (1989), *Ap.J.*, **341**, 129.
4. Casoli, F., *et al.* (1998), *Astr.Ap.*, **331**, 451.
5. Cox, P., Krugel, E., and Mezger, P. (1986), *Astr.Ap.*, **155**, 380.
6. Devereux, N., and Young, J. (1990a), *Ap.J.(Letters)*, **350**, L25.
7. Devereux, N., and Young, J. (1990b), *Ap.J.*, **359**, 42.
8. Devereux, N., and Young, J. (1991), *Ap.J.*, **371**, 515.
9. Devereux, N., and Young, J. (1992), *A.J.*, **103**, 1536.
10. Devereux, N., and Young, J. (1993), *A.J.*, **106**, 948.
11. Draine, B.T. and Lee, H.M. (1984), *Ap.J.*, **285**, 89.
12. Haas, M., *et al.* (1998), *Astr.Ap.*, **338**, L33.
13. Habing, H., *et al.* (1984), *Ap.J.*, **278**, L59.
14. Helou, G., *et al.* (1985), *Ap.J.(Letters)*, **305**, 15.

15. Heyer, M.H., *et al.* (1989), *Ap.J.*, **346**, 220.
16. Hildebrand, R.H. (1983), *Quart.J.R.A.S.*, **24**, 267.
17. Hunter, D., *et al.* (1989), *Ap.J.*, **341**, 697.
18. Joseph, R.D., and Wright, G. (1985), *M.N.R.A.S.*, **214**, 87.
19. Jura, M. (1982), *Ap.J.*, **254**, 70.
20. Krugel, E., *et al.* (1998), *Astr.Ap.*, **331**, L9.
21. Lees, J., Knapp, G., Rupen, M., Phillips, T. (1991), *Ap.J.*, **379**, 177.
22. Lonsdale, C., and Helou, G. (1987), *Ap.J.*, **314**, 513.
23. Maloney, P., and Black, J. (1988), *Ap.J.*, **325**, 389.
24. Mezger, P., Mathis, J., and Panagia, N. (1982), *Astr.Ap.*, **105**, 372.
25. Noguchi, N., and Ishibashi, S. (1986), *M.N.R.A.S.*, **219**, 305.
26. Rengarajan, T.N., and Verma, R.P. (1986), *Astr.Ap.*, **165**, 300.
27. Rice, W., *et al.* (1990), *Ap.J.*, **358**, 418.
28. Rubin, V., *et al.* (1985), *Ap.J.*, **289**, 81.
29. Sage, L. (1993), *Astr.Ap.*, **272**, 123.
30. Sanders, D.B., and Mirabel, I.F. (1985), *Ap.J.(Letters)*, **298**, L31.
31. Sanders, D.B., Scoville, N., and Soifer, B.T. (1991), *Ap.J.*, **370**, 158.
32. Sauvage, M., and Thuan, T. (1992), *Ap.J.(Letters)*, **396**, L69.
33. Scoville, N., and Good, J. (1989), *Ap.J.*, **339**, 149.
34. Smith, B., *et al.* (1994), *Ap.J.*, **425**, 1994.
35. Smith, J. (1982), *Ap.J.*, **261**, 463.
36. Solomon, P.M., and Sage, L. (1988), *Ap.J.*, **334**, 613.
37. Stark, A., *et al.* (1986), *Ap.J.*, **310**, 660.
38. Stark, A., *et al.* (1989), *Ap.J.*, **337**, 650.
39. Telesco, C., and Harper, D. (1980), *Ap.J.*, **235**, 392.
40. Thronson, H., *et al.* (1989), *Ap.J.*, **344**, 747.
41. Tinney, C., *et al.* (1990), *Ap.J.*, **362**, 473.
42. Wilson, C. (1995), *Ap.J.(Letters)*, **448**, L97.
43. Wynn-Williams, C.G., and Becklin, E. (1974), *P.A.S.P.*, **86**, 5.
44. Young, J. (1999), *Ap.J.(Letters)*, **514**, L87.
45. Young, J., *et al.* (1996), *A.J.*, **112**, 1903.
46. Young, J., *et al.* (1986a), *Ap.J.Letters*, **311**, L17.
47. Young, J., Kenney, J., *et al.* (1984), *Ap.J.(Letters)*, **287**, L65.
48. Young, J., and Knezek, P. (1989), *Ap.J.Letters*, **347**, L55.
49. Young, J., Schloerb, F., *et al.* (1986b), *Ap.J.*, **304**, 443.
50. Young, J., and Scoville, N. (1991), *Ann.Rev.Astr.Ap.*, **29**, 581.
51. Young, J., Xie, S., *et al.* (1989), *Ap.J.Supp.*, **70**, 699.
52. Young, J., *et al.* (1995), *Ap.J.Supp.*, **98**, 219.

THE SCUBA LOCAL UNIVERSE GALAXY SURVEY - A LOW REDSHIFT BENCHMARK FOR THE DEEP SUBMM SURVEY

D.L. CLEMENTS, L. DUNNE, S.A. EALES

*Dept. Physics and Astronomy, Cardiff University, PO Box 913,
Cardiff, CF24 3YB, UK*

The SCUBA Local Universe Galaxy Survey (SLUGS)[4] is the first statistical survey of the properties of local galaxies in the submm. The first part of this survey consisted of observations of a complete sample of galaxies selected from the IRAS Bright Galaxy Survey (BGS). This allows us to produce the first $850\mu m$ luminosity function for local galaxies, calculate dust temperatures, spectral energy distributions and the dust mass function, and to compare submm properties with results at other wavelengths. The survey is now being extended to both a larger sample of IRAS galaxies, including more high luminosity objects, and to a complete sample of optically selected galaxies. Since little was previously known about the submm properties of local galaxies, the SLUGS project provides an important baseline against which objects found in deep submm surveys may be compared. Our local luminosity function, for example, is significantly different from simple extrapolations to the $60\mu m$ IRAS luminosity function which have often been used in the interpretation of deep submm survey data.

1 Why a Local Submm Galaxy Survey?

The introduction of submm bolometer arrays has brought great strides in our understanding of the high redshift universe. We have begun to find the sources that produce the Cosmic Infrared Background (CIB),[8] and which seem to be the site of copious amounts of star formation.[6,7] With all this progress at high redshift it would be easy to assume that low redshift sources are well defined at submm wavelengths. Sadly, this is far from true. We currently do not even know the dust temperature and spectral energy distribution for typical local galaxies, let alone its spatial distribution and its relation to other material in the emitting galaxy. Only a handful of local galaxies have been thoroughly studied in the submm (*e.g.* Alton et al. 2000), and many of these are special in some way. Studies of larger samples of galaxies in the submm, are scarcer still. Some attempts were made with early single element bolometers (*e.g.* Clements et al., 1993), but we have had to await bolometer arrays to avoid many uncertainties in the interpretation of such data. With bolometer arrays and statistically complete samples we can now attempt to determine the local submm luminosity function, dust mass function, and spectral energy distributions of galaxies. This is the aim of the SLUGS project.[4]

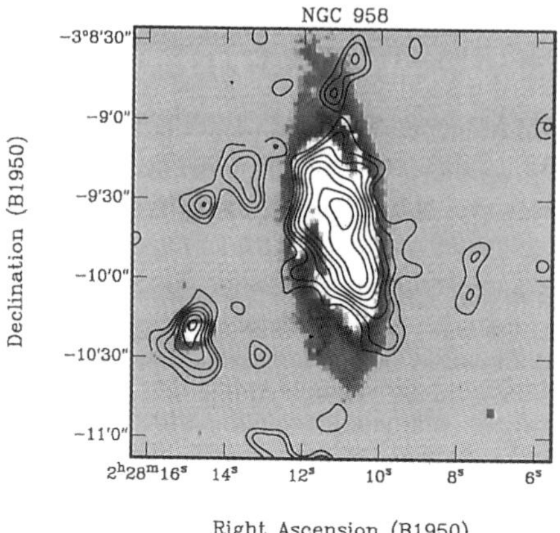

Figure 1. SCUBA image of NGC958 (contours) overlaid on optical DSS image.

2 Samples and Observations

The observations for SLUGS were all made with SCUBA on the JCMT using jiggle-map mode. This observing mode provides a circular field-of-view $\sim$2.2 arcmin across, which provides a limit on the size of objects that we can observe. The first complete sample of galaxies observed by SLUGS were selected from the IRAS BGS [9], a complete 60μm flux limited sample with $F_{60} > 5.24$ Jy. We select a subsample from this with $-10 < \delta(\text{deg.}) < 50$ and velocity >1900 kms^{-1}. This selection covers an area of 10400 deg^2 and contains 104 galaxies. The bulk of the results discussed in this paper are derived from this sample. An example image from the survey is shown in Fig. 1.

The point of selecting sources from the IRAS database is to define a statistical sample of objects likely to be bright in the submm. We can then apply accessible volume techniques to determine, for example, the luminosity function of the population. Such techniques are effective when all possible classes of source are represented in the selected sample. However, it is possible that some sources that are luminous in the submm are not prominent in the far-IR, where the IRAS selection is made. This could result, for example, from a population of galaxies containing dust colder than $\sim$30 K, suppressing the far-IR emission at 60μm but still emitting strongly in the submm. We are thus in the process of extending the SLUGS survey to a sample of optically selected galaxies. Provisional results from the incomplete optically selected sample so far observed does indeed indicate a lower dust temperature than for the IRAS selected sample.[4] These observations are continuing.

3 Results: Luminosity Function

Fig. 2 shows the 850μm luminosity function calculated for the IRAS selected SLUGS galaxies. This is significantly different from luminosity functions derived from single temperature extrapolations from the IRAS luminosity function, as used to interpret some of the deep submm survey data. These typically under-predict the number density of lower luminosity objects since there is a 60μm flux – temperature correlation that is absent at 850μm. Lower 850μm luminosity objects are thus under-represented in the 60μm surveys used for the extrapolation. The effect of this on the interpretation of deep SCUBA surveys is to over-estimate the evolution required from the low redshift population to match the counts seen in the deep, supposedly, high redshift population.

4 Calibrating the Radio/Submm Redshift Relation

Carilli & Yun (1999) have suggested the use of the ratio of submm to radio emission as a redshift indicator for the deep surveys. Much of this technique was originally based on theoretical considerations, and lacked a low redshift calibration set, made up of a complete sample of submm observed objects. We have now used the SLUGS survey galaxies to provide such a calibration, and this is shown in Fig. 2 together with deep survey objects with spectroscopic redshifts.[5] The one source that lies away from our relation is already known to be an AGN and thus will have an underestimated redshift using this method. The overall effect of using the SLUGS calibration to this relation is to reduce the derived redshifts for the deep sources somewhat.

5 Conclusions

The SLUGS survey has produced the first statistically complete survey of the submm properties of local galaxies. As such it as an important benchmark for the understanding of deep submm surveys. The SLUGS survey implies that a greater range of uncertainty exists in the properties of the faint submm population than previously assumed, and that a wide range of evolutionary histories and roles are consistent with the data.[6,4,5] To remove some of these uncertainties two things are needed: more direct measurements of redshifts for

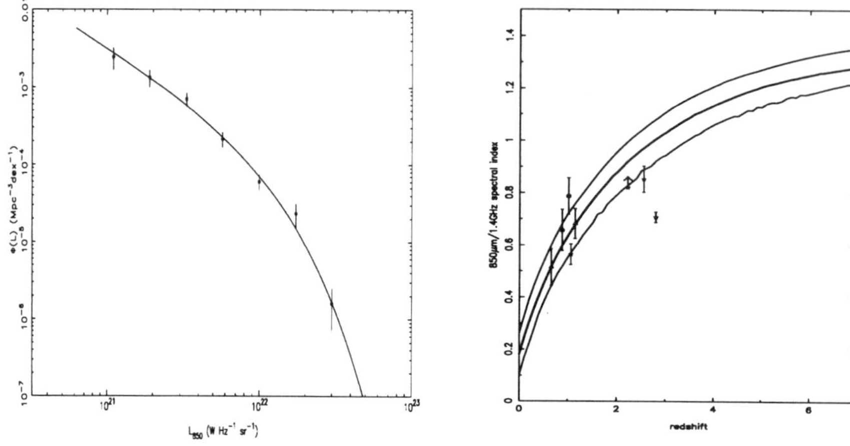

Figure 2. Left: SLUGS-derived local 850μm Luminosity Function; Right: SLUGS calibrated radio/submm redshift relation. For the radio/submm redshift relation, the lines show the median SLUGS data, together with 95th percentile points on either side, allow uncertainties to be calculated. Data points are for deep SCUBA sources with spectroscopically measured redshifts.

the deep populations, and the determination of the temperature, and thus the luminosity, for these same objects. Hopefully 8-m class optical/IR telescopes and SIRTF and SOFIA will allow this data to be collected in the next few years.

References

1. Alton, P. *et al*, *A&A* **356**, 795 (2000).
2. Carilli, C.L. & Yun, M.S., *ApJ* **513**, L13 (1999).
3. Clements, D.L. *et al*, *MNRAS* **261**, 299 (1993).
4. Dunne, L. *et al*, *MNRAS* **315**, 115 (2000a).
5. Dunne, L. *et al*, *MNRAS*, in press (2000b).
6. Eales, S.A. *et al*, *AJ*, in press (2000).
7. Ivison, R.J. *et al*, *MNRAS* **315**, 209 (2000).
8. Puget, J-L. *et al*, *A&A* **308**, L5 (1996).
9. Soifer, B.T. *et al*, *AJ* **98**, 766 (1993).

THE HIGH-RESOLUTION CO SURVEY OF M 31

N. NEININGER

Radioastr. Inst. Univ. Bonn, Auf dem Hügel 71, D-53121 Bonn, Germany
E-mail: nneini@astro.uni-bonn.de

We are about to finish a fully sampled high resolution survey of the molecular gas in the nearby spiral M 31. It covers the whole disk ($\sim$ 1 sq degree) at 23″ resolution. The good stability of the receivers resulted in a very uniform base level. A typical noise figure in the final spectra is 30 mK rms. The molecular gas in M 31 is distributed in long, narrow spiral arm segments that show a much higher arm/interarm contrast than the atomic gas. Significant molecular emission is found between radii of about 4 kpc to 18 kpc. On average, the ratio of atomic to molecular gas rises with radius. Close to the nucleus a special observing technique was necessary to detect any molecular gas – at a level 10 times lower than in the disk. Strong large-scale streaming motions are absent – they seem to be at the 10 km s^{-1} level which requires a detailed analysis to disentangle them from the random motions of the molecular clouds. However, several regions in the galaxy show broad and even multiple-component spectra with separations of up to 50 km s^{-1} generated by local effects.

1   The Background and the Technique

1.1 · Local templates

The study of the early universe or, more generally, distant objects needs to be supported by detailed studies of local objects to derive templates for the analysis of the necessarily limited amount of information available. Typical choices for active systems are M 82, NGC 253, Arp 220 or Mrk 273 – for "normal" spirals usual templates are objects like M 31, M 51 or M 81. We set out to study the global properties of the molecular gas as a whole together with the detailed conditions in normal star forming regions – with the highest possible spatial resolution and sensitivity. The obvious choice is the nearest external spiral galaxy, M 31.

Of course, it was already on the target list soon after the first detection of extra-terrestrial CO, but the combination of low activity and large angular size made the observations very difficult. Therefore, only a few small regions have been observed so far at angular resolutions below 1′; the first survey of the ^{12}CO has been completed only a few years ago by Dame et al.[2], with a beam size of 8.′7.

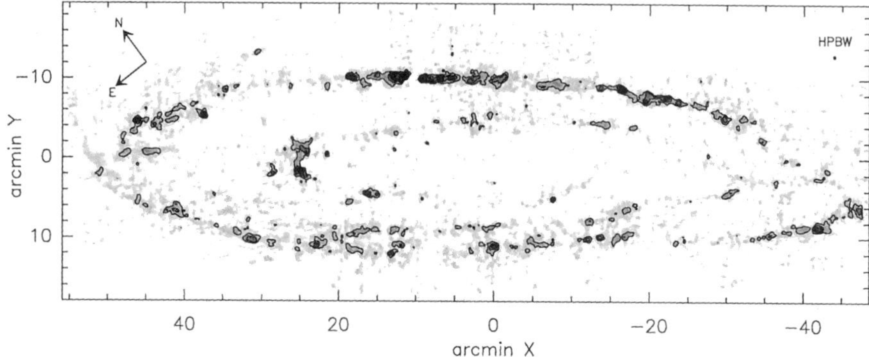

Figure 1. The main ridges of the CO emission of M 31. Contours start at 5σ of the integrated flux. Many cloud complexes with narrow line widths $(4\ldots10\ \mathrm{km\,s^{-1}})$ are missing in this representation. The dot in the upper right corner indicates the beam size; the orientation is given in the upper left corner.

1.2 *Observations*

We started a new survey at twenty times higher angular resolution ($23''$) in November 1995 using the IRAM 30-m telescope on Pico Veleta. In order to cover the large area in a reasonable time we applied a technique developed for centimetric continuum observations, the "On-The-Fly" method: The telescope is scanned across the source at a constant speed and spectra are taken at a rate high enough to ensure a proper sampling of the beam. Tests had revealed that the receiver stability allows scans of a duration of several minutes between two reference measurements of the blank sky. The typical line profiles in M 31 are much smaller than the typical backend bandpass of 256 MHz ($\simeq 660\,\mathrm{km\,s^{-1}}$) so that a large part of the remaining atmospherically induced variations is removed already with the usual spectral baseline fit. For the remaining "scanning noise" we applied a dedicated filtering algorithm [3], adapted to spectroscopic data [4]; this proved to be very efficient and reliable. A detailed description of the observational technique is given elsewhere [7].

2 Properties of the molecular gas

2.1 *The CO distribution*

Very early it became evident that the molecular gas in M 31 is distributed in narrow, almost filamentary structures (see Fig. 1). The width of the CO emission in the radial direction is of the order of one minute of arc. In tangential direction, some prominent arm segments reach a length of about 20′, but in general the whole distribution appears very patchy – although clearly organized along a spiral pattern[6]. The "arm" and the "interarm" regions are well defined and the contrast reaches values of 10:1. This is about three times higher than the contrast determined at the same places in the distribution of the neutral atomic hydrogen (HI) in a survey of similar angular resolution[1].

The general distribution of the molecular gas, however, resembles that of the HI: unlike most other spiral galaxies the central region of M 31 is devoid of any neutral gas component. We detected significant emission above a level of 100 mK only in a "ring" between radii of 4 kpc and 18 kpc. Thus the inner boundary is about the same for the atomic and the molecular gas. Of course, the ratio does not remain constant: in the inner parts the molecular gas dominates and towards larger radii the atomic gas becomes the major constituent.

A special effort was made to obtain information about the distribution of the molecular gas close to the center[5]. At a few hundred parsecs distance we found indeed CO emission in the (1-0) and the (2-1) transition, but with very low intensity of about 20 mK only. The line ratio and other properties turned out to be rather "normal", suggesting a temperature of > 10 K and general properties like local Galactic clouds.

2.2 Kinematics

From earlier observations of a small region at an angular resolution of 1″.7 rather large streaming motions of the order of $30\,\mathrm{km\,s^{-1}}$ had been derived in the north-east[9]. Our survey shows however no signs of such systematic velocity deviations. There are indications of weak effects in the range of $10 \ldots 15$ $\mathrm{km\,s^{-1}}$, but this is of the order of the random motions of the individual molecular clouds and the intrinsic line widths. A detailed kinematical modeling and analysis is needed to clearly separate the individual contributions.

Spectra with multiple components and a large separation between them have however been found in many *individual* cloud complexes. The line splitting reaches values of up to 50 $\mathrm{km\,s^{-1}}$, but is limited to regions of at most a few arcminutes across. Usually, these spectra show large variations on small angular scales as well, which makes these regions ideal targets for follow-up interferometric observations[8].

The detailed nature of these flows is not yet understood, but in contrast to Galactic studies we can clearly determine their location and compare them with other tracers because of the "outside view", the inclination of M 31, the high spatial resolution, and the negligible distance uncertainty.

3 First results

A comparison of the total gas content with the optical extinction in the middle of the "emission ring" indicates that in M 31 the CO is a good tracer of the molecular hydrogen[6]. A more detailed study is however needed to check for possible radial variations or local anomalies. The small width of the CO emission ridges and the high arm/interarm contrast suggest that the molecular clouds are indeed short-lived objects, even in the absence of strong streaming motions or an intense interstellar radiation field. The deviations from a regular spiral structure (in particular the bifurcation in the east and the disturbed structure in the south-west) as well as the missing CO in the inner part of M 31 may indicate a more troubled past of this quiescent galaxy.

Acknowledgments

The survey work is done by a collaboration of several people and in particular M. Guélin, P. Hoernes, R. Lucas, C. Nieten, H. Ungerechts and R. Wielebinski (in alphabetical order) have made significant contributions. The indispensable framework at the Pico Veleta Observatory was prepared and maintained by the receiver engineers in Grenoble and Granada and by the computer and data acquisition group at the telescope.

References

1. E. Brinks, W.W. Shane, *Astron. & Astrophys. Suppl. Ser.* **55**, 179 (1984)
2. T.M. Dame, E. Koper, F.P. Israel, P. Thaddeus, *Astrophys. J.* **418**, 730 (1993).
3. D.T. Emerson, R. Gräve, *Astron. & Astrophys.* **190**, 353 (1988)
4. P. Hoernes, *PhD Thesis*, University of Bonn (1997)
5. A.-L. Melchior, F. Viallefond, M. Guélin, N. Neininger, MNRAS **312**, L 29 (2000)
6. N. Neininger, M. Guélin, H. Ungerechts, R. Lucas, R. Wielebinski, *Nature* **395**, 871 (1998)
7. N. Neininger, in *Imaging at Radio through Submillimeter Wavelengths*, ed. J. Mangum, (PASP Conference Series, 2000), p. 52.
8. N. Neininger, in *The Interstellar Medium in M 31 and M 33, 232. WE-Heraeus Seminar*, eds. E.M. Berkhuijsen et al., (2000, in press)
9. B.S. Ryden, A.A. Stark, *Astrophys. J.* **305**, 823 (1986)

3. Facilities and Instrumentation

MILLIMETER OBSERVATIONS USING THE GREEN BANK TELESCOPE

SIMON R. DICKER, MARK J. DEVLIN, JASON L. PUCHALLA, JEFF KLEIN

University of Pennsylvania, Philadelphia, PA 19104, USA

The new Green Bank telescope (GBT) is expected to operate at frequencies up to 100 GHz. The University of Pennsylvania operates a seven element bolometer array which, with minor modifications, could be used to make observations on the GBT. This talk contained an analysis of the sensitivities which would be obtained using this instrument and gave examples of observations which could be made.

1 Introduction.

The 100 meter Green Bank Telescope (GBT) was designed as a replacement for the 300 ft telescope which collapsed in 1988. Built at the NRAO site in Green Bank, West Virginia and protected by a radio-transmitter-free zone, this telescope is the world's largest fully steerable telescope.[1]

The GBT has a primary surface made of 2004 active panels, each of which can be positioned to better than 25μm using an actuator at its corners. When the active surface is fully operational these actuators will be driven by a laser metrology system to give a primary surface RMS better than 0.24 mm at all elevation angles.[1] This yields a 90 GHz surface efficiency of 36%. The enormous surface area of the GBT means that even with this relatively modest surface efficiency, the GBT will have a 90 GHz gain of 0.846 K/Jy making it the most sensitive 90 GHz telescope in existence.

The University of Pennsylvania plans to modify an existing array of 140 and 90 GHz bolometers to an array of seven 90 GHz bolometers and to use this array on the GBT. The Penn Array is a fast and cost effective way of getting the first 90 GHz science from the GBT and will provide valuable 90 GHz engineering data to guide the design of future high-frequency instruments.

2 90 GHz Science With The GBT.

Table 1 shows expected noise levels for the Penn Array on the GBT. The atmosphere contributes only 10% to these values. With the active surface fully operational, noise levels will be as low as 1.3 mJy $\mathrm{Hz}^{-1/2}$. This section gives a few examples of the science that could achieved with these noise levels and the 8 arcsec beam size of the Penn Array on the GBT.

Table 1. The 90 GHz sensitivity to point sources for the Penn Array on the GBT in typical observing conditions, assuming a sky opacity of $\tau = 0.10$ and a sky temperature of 28 K. For Comparison, the IRAM 30 m telescope currently has a sensitivity of 12 mJy $Hz^{-1/2}$. The RMS surface accuracies given are for the different stages of the GBT project. The correspond (from the left) to observing at 44 deg. elevation with the active surface off, using the active surface controlled by a model of the telescope and using the active surface controlled by the laser metrology system.

GBT surface accuracy RMS	0.48 mm	0.36 mm	0.24 mm
Receiver noise (mJy $Hz^{-1/2}$)	23.6	4.3	1.2
Atmospheric noise (mJy $Hz^{-1/2}$)	6.6	1.2	0.3
TOTAL (mJy $Hz^{-1/2}$)	25.8	4.7	1.3

2.1 High Redshift Galaxies

By early 2001, the SCUBA camera on the JCMT will have completed the first 850 mm surveys with 3-sigma depths in the range 1.5–8 mJy.[2,3,4] Many of the objects in this survey are expected to be galaxies at high redshift. Due to their faintness, each redshift measurement currently requires up to 6 hours of integration time on an 8 m-class optical telescope. Taking into account the spectrum of the submillimeter sources, the Penn Array will produce maps with the same signal to noise ratio as the SCUBA maps using integration times of a few hundred seconds/pixel. Such observations would provide photometric redshifts for many more SCUBA sources than have currently been measured.

2.2 Circumstellar Disks

Using 1.3 mm continuum maps, Beckwith et al. detected circumstellar disks in $\sim$40% of the pre-mainsequence stars they observed.[7] By combining their data with radio and IR observations, information on how the density and temperature of these disks changed with radius was obtained. Evidence of particle growth was also observed. Beckwith's maps had a resolution of 11 arcsec and required 200 secs of integration per beam to reach noise levels of 5 mJy. With one second of integration per beam the Penn Array would reach better noise levels allowing deeper and wider surveys of pre-mainsequence stars to be made. Such sensitive measurement would enable progress to be made on the theories of the formation of planets.

2.3 *The Sunyaev-Zel'dovich Effect*

Measurements of the Sunyaev-Zel'dovich (SZ) effect in giant clusters of galaxies could also be made using the Penn Array. Typical SZ clusters are ~ 1 arcmin in size and the magnitude of the SZ effect at 90 GHz is 0.2 mK. In good weather a 1 sq. arcmin could be mapped to an r.m.s. of 0.06 mK in 5–10 hours. Although the SZ effect has been detected before[8] using beam sizes comparable to or larger than the clusters, the 8 arcsec beam size of the GBT at 90 GHz would be able to resolve the centers of clusters providing critical information on the distribution of hot gas in clusters.

3 The Penn Array Receiver.

The Penn Array Receiver was originally designed to observe the Cosmic Microwave Background (CMB). The array configuration for CMB observations consisted of five 140 GHz bolometers, arranged in a close-packed array, two 90 GHz feeds on either side and one dark channel. Electronics exists to operate up to eight spider bolometers[5] simultaneously. These bolometers have NEPs $< 10^{-18}$ W/Hz$^{1/2}$.

The bolometers, the corrugated horn feeds used to couple them to the telescope, and the ceramic load resistors used to bias them are cooled to < 0.3 K using a closed cycle ^{3}He refrigerator. The liquid ^{3}He in the refrigerator can be regenerated automatically or remotely using a continuously pumped ^{4}He reservoir that is fed by a capillary from the main ^{4}He dewar.[6] Regeneration requires ~ 3 hours. Cryogens need to be replenished every 5 days which is compatible with the maintenance schedule of the GBT. The ^{3}He refrigerator lifetime depends on load but is typically 4 days. All of this cryogenic equipment and electronics *exists and has been tested*. All that will be required to adapt this receiver to work on the GBT will be the design and mounting of new feeds, the updating of the electronics and the integration of the control system into the GBT.

4 The Atmosphere at Green Bank.

Telescopes with very sensitive receivers are often limited in sensitivity by the atmosphere above the site. Consequently an accurate evaluation of the atmospheric opacity above the GBT site is vital. NRAO operates an 86 GHz tipper, located approximately 1 mile form the GBT site (`http://www.gb.nrao.edu`). Models of the atmosphere predict that this is a good measure of the opacity at 90 GHz. The cumulative statistics from data taken in March 1999 show

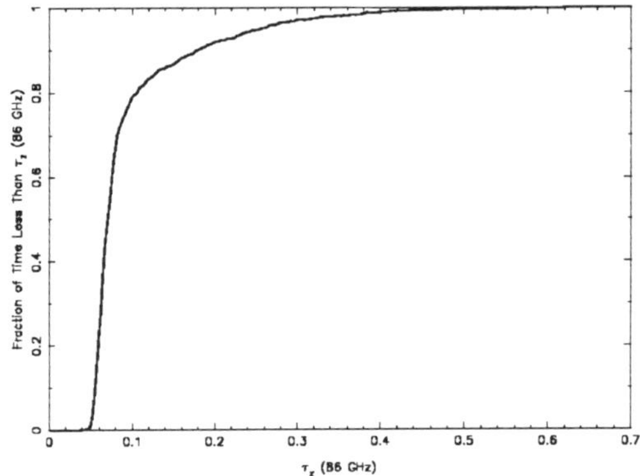

Figure 1. Atmospheric 86 GHz opacity at Green Bank in March 1999.

that for 20% of the time the opacity was below 0.06 (a sky temperature of 15 K) and for 80% of the time the opacity was below 0.1 (a sky temperature of 28 K). These values demonstrate that during the winter, the atmosphere at Green Bank is of suitable quality for 90 GHz observations.

5 Conclusions.

On a timescale of 2–3 years the Penn Array should be operating on the GBT as a user instrument with seven 90 GHz pixels each with a resolution of 8 arcsec. It will have sensitivities up to 1.3 mJy Hz$^{-1/2}$ compared $\sim$ 12 mJy Hz$^{-1/2}$ for existing experiments. The Penn Array receiver will also provide important data for the design of a more complex 90 GHz camera for the GBT.

References

1. Lockman, F.J., *Proc. SPIE* **3357**, 656 (1998).
2. Smail, I., Ivison, R.J., Blain, A.W., *ApJ* **490**, L5 (1997).
3. Hughes, D.H. *et al, Nature* **394**, 241 (1998).
4. Lilly, S. *et al, ApJ* **518**, 641L (1999).
5. Bock, J. *et al, Proc. SPIE* **3357**, 297 (1998).
6. Puchalla, J.P. *et al*, in preparation
7. Beckwith, S.V.W. *et al, AJ* **99**, 924 (1990).
8. Komatsu, E., *ApJ* **516**, L1 (1999).

SOME PROSPECTS FOR HIGH REDSHIFT GALAXY OBSERVATIONS WITH THE SUBMILLIMETER ARRAY

D.J. WILNER

Harvard-Smithsonian Center for Astrophysics,
60 Garden St., Cambridge, MA 02138
E-mail: dwilner@cfa.harvard.edu

The Submillimeter Array, being built near the summit of Mauna Kea by the Smithsonian Astrophysical Observatory and the Institute of Astronomy and Astrophysics (Taiwan), has the potential to make unique observations of the newly recognized population of luminous high redshift dusty galaxies. Prospects for submillimeter continuum observations include (1) accurate position determinations for improved associations at optical/infrared/radio wavelengths, (2) deep integrations that probe beyond the confusion limit of single dish telescopes, and (3) imaging the detailed morphologies of the stronger sources at kiloparsec size scales.

1 The SMA Project

The Submillimeter Array (SMA) is a collaborative project of the Smithsonian Astrophysical Observatory and the Institute of Astronomy and Astrophysics of the Academia Sinica. A brief history of the SMA project and a detailed technical description of the array can be found in a recent article by James Moran.[3] The SMA will consist of eight 6 meter diameter antennas sited near the summit of Mauna Kea. As of August 2000, three antennas have been deployed in Hawaii. Each antenna will be equipped with up to eight receivers covering the atmospheric windows from 1.4 to 0.3 millimeters. The SMA is an interferometer with maximum resolution about $0\rlap{.}''1$ at the shortest operating wavelength. The nominal rms continuum sensitivity of the full array is expected to be about 1.0 mJy in the 850 μm atmospheric window after 8 hours observing with one polarization (assuming 1 mm of precipitable water).

When the SMA was originally proposed– more than fifteen years ago– the scientific justification did not consider observations of cosmologically distant sources.[2] As is now well known, there exists a large population of luminous dusty galaxies at high redshifts that are detectable at submillimeter wavelengths. The steep spectral index longward of the far-infrared emission peak results in a negative K-correction that effectively offsets the dimming due to distance, resulting in almost constant emission for a wide range of redshifts. Because the SMA was designed with collecting area comparable to the largest existing single dish submillimeter telescopes, the sensitivity of the full array will be adequate to access all sources that are well detected with e.g. the

James Clerk Maxwell Telescope (JCMT) and remain unresolved, including the recently recognized high redshift sources.

2 Some Prospects for the SMA

2.1 Accurate Positions

The SMA should routinely provide sub-arcsecond position information for the high redshift sources discovered with bolometer arrays like *SCUBA* on the JCMT. Accurate position determinations will aid the current painstaking efforts to identify counterparts at optical, near-infrared, and radio wavelengths, if present, as the poor resolution of existing submillimeter data often allows several candidates. Reliable associations will help to establish the basic physical properties of these systems and their redshift distribution.

2.2 Deep Surveys

The SMA may make deep integrations into the regime where the existing single dish telescopes become confusion limited. There is good evidence that confusion becomes problematic at the 2 mJy level at 850 μm with *SCUBA*.[4] Observations that probe lower flux levels may allow for additional associations with classes of objects selected at other wavelengths, for example with optical Lyman break galaxies. These fainter submillimeter sources may prove important for distinguishing aspects of competing star formation histories.

To illustrate the potential of a deep survey with the SMA, we have simulated observations of high declination blank fields. We populate the sky following the prescriptions of Blain et al. (1999) who present families of models for the evolution of dusty galaxies consistent with available data at far-infrared and submillimeter wavelengths, including the evolution of 60 μm counts from *IRAS* at low redshift, the 175 μm and 850 μm counts from *ISO* and *SCUBA*, and the diffuse background spectrum from *COBE*.[1] The models have three main ingredients: a representative spectral shape for individual galaxies, a luminosity function (based on 60 μm measurements in the local universe), and pure luminosity evolution that is parameterized by the low redshift slope of an evolution function, a redshift beyond which evolution slows down, and a redshift at which the first galaxies appear.

Figure 1 shows two views of the model sky based on the "Anvil-5" evolution parameters of Blain et al. (1999). Each galaxy is assumed to fill a 1″ spot on the sky, and the positions are distributed randomly (i.e. there is no clustering due to large scale structure). The upper panel shows the view of a region 256″ × 256″ with *SCUBA* resolution and no noise. The source near the

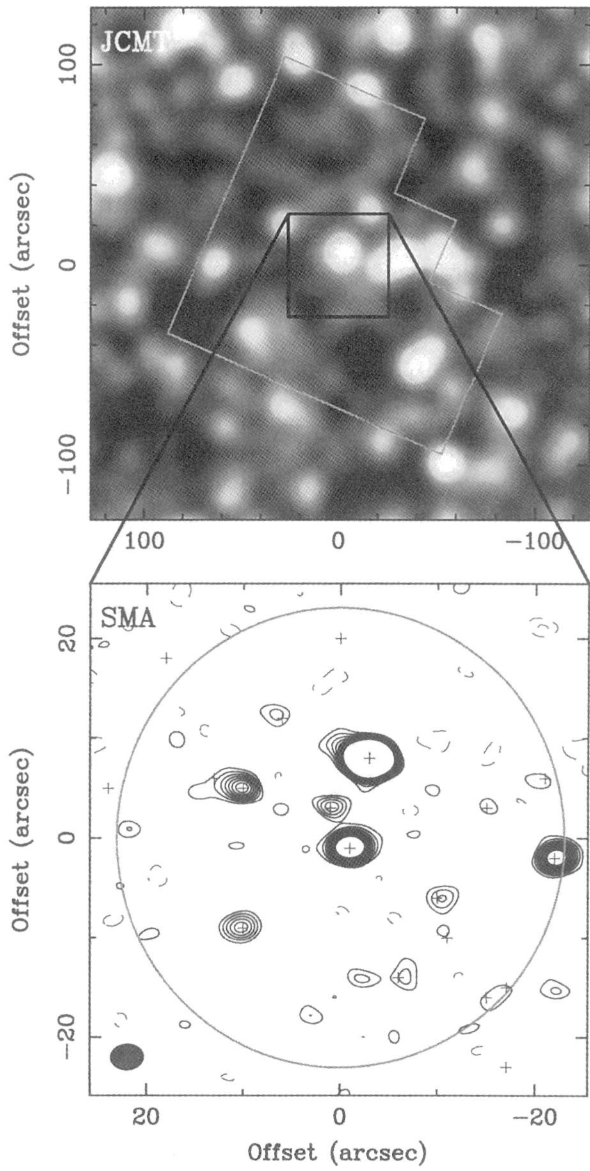

Figure 1. Simulated 850 μm sky at $14''$ resolution (upper panel), comparable to an observation made with *SCUBA* on the JCMT, and a deep field observed with the SMA with $\sim 3''$ resolution (lower panel). The lowest contour level is 0.2 mJy and the step is 0.1 mJy. The circle indicates the 30% response level of the SMA primary beam.

box center has an apparent flux of 8.5 mJy. The lower panel shows a simulated deep SMA integration, which resolves the emission into many individual galaxies. The simulation assumes 13 tracks in each of the two compact SMA configurations, observed with two polarizations, with 75% of the time spent on the target field (total time $\sim$ 200 hours). The 850 μm zenith sky opacity is taken to be 0.18, the first quartile value from extensive site monitoring in the 1990's. The synthesized beam size is $3\rlap{.}''4 \times 2\rlap{.}''5$, and the rms noise is about 0.1 mJy. Crosses mark all of the sources in the model with fluxes greater than 0.3 mJy. Nearly all of these faint sources within the $35''$ half power primary beam width of the SMA can be readily identified in the image.

2.3 Resolving Structure in Dusty Galaxies

The stronger high redshift galaxies will be amenable to imaging at subarcsecond resolution with the SMA, which will probe kiloparsec size scales. Especially important for this work are provisions underway to include the JCMT and CSO as occasional elements in the array. The addition of these two large telescopes will double the point source sensitivity and increase the resolution by about 50%. These enhancements are important because even the stronger high redshift objects will be difficult to image in detail with the SMA alone. Images made with the SMA including the JCMT and CSO will have sub-mJy sensitivity in 8 hours and $0\rlap{.}''2$ resolution under good atmospheric conditions. Such observations have the potential to reveal a wealth of small scale structure, e.g. the locations and sizes of obscured starbursts and the morphologies of merging systems, and they will allow for detailed comparisons with sub-arcsecond data obtained at other wavelengths.

Acknowledgments

The SMA involves a dedicated staff in Massachusetts, Hawaii and Taiwan.

References

1. A.W. Blain, I. Smail, R.J. Ivison, J.-P. Kneib, MNRAS, **302**, 632 (1999).
2. J.M. Moran *et al.*, A Submillimeter Wavelength Telescope Array: Scientific, Technical, and Strategic Issues, Smithsonian Astrophysical Observatory Special Report (1984).
3. James M. Moran, Proc. SPIE Vol. 3357, Advanced Technology MMW, Radio, and Terahertz Telescopes, ed. Thomas G. Phillips, p. 208 (1998).
4. J.A. Peacock *et al.* astro-ph/9912231 (2000).

BLAST – A BALLOON-BORNE LARGE APERTURE SUBMILLIMETRE TELESCOPE

MARK J. DEVLIN* FOR THE BLAST COLLABORATION[†]

*Department of Physics and Astronomy, University of Pennsylvania,
Philadelphia, PA 19104, USA
E-mail: devlin@physics.upenn.edu*

A new generation of sub-orbital platforms will be operational in the next few years. These new telescopes will operate from airborne and balloon-borne platforms where the atmosphere is transparent enough to allow sensitive measurements to be made in the submillimeter bands. The telescopes will take advantage of state-of-the-art instrumentation including large format bolometer arrays and spectrometers. Other papers in this volume will deal specifically with the potential of these bands. In this paper we review the capabilities of the BLAST balloon-borne telescope.

1 Introduction to BLAST

The 'Balloon-borne Large-Aperture Sub-millimeter Telescope' (BLAST) incorporates a $2.0\,m$ mirror with plans to increase to a $2.5\,m$ mirror. The telescope will operate on a Long Duration Balloon (LDB) platform with large format bolometer arrays at 250, 350 and $500\,\mu m$. BLAST will address some of the most important Galactic and cosmological questions regarding the formation and evolution of stars, galaxies and clusters.

---

[a] Alfred P. Sloan Fellow
[b] The BLAST collaboration includes:
Peter Ade - Queen Mary and Westfield College, London, UK
James Bock - JPL, Pasedena, CA, USA
Paolo DeBernardis - University of Rome, Rome, Italy
Mark Devlin - University of Pennsylvania, Philadelphia, PA, USA
Joshua Gundersen - Princeton University, Princeton, NJ, USA
Mark Halpern - University of British Columbia, Vancouver, CA
David Hughes - Instituto Nacional de Astrofisica, Optica y Electronica, Puebla, Mexico
Jeff Klein - University of Pennsylvania, Philadelphia, PA
Phillip Mauskopf - Cardiff University, Cardiff, UK
Silvia Masi - University of Rome, Rome, Italy
Barth Netterfield - University of Toronto - Toronto, Canada
Luca Olmi - University of Massachusetts, Amherst, MA, USA
Lyman Page - Princeton University, Princeton, NJ, USA
Douglas Scott - University of British Columbia, Vancouver, Canada
Gregory Tucker - Brown University, Providence, RI, USA

The primary advantage of BLAST over existing and planned ground-based and airborne sub-mm telescopes is the dramatically increased atmospheric transmission at balloon altitudes (Fig.1.). This results in greatly enhanced instrumental sensitivity at wavelengths $\leq 500\,\mu m$. BLAST complements the FIRST satellite by overlapping the FIRST frequency coverage as well as having the ability to test new technologies for future space-based missions. BLAST is the only facility capable of providing sufficient field of view, sensitivity, and integration time to conduct follow-up of SIRTF/MIPS surveys at 200-400 μm.

2 Science Goals of BLAST

BLAST will be the first balloon-borne telescope to take advantage of the bolometric focal plane arrays being developed for FIRST. It will be capable of probing the sub-mm with high spatial resolution and sensitivity providing the opportunity to conduct unique Galactic and extragalactic surveys. Compared to the pioneering flights of PRONAOS[13], BLAST will have an advantage of > 100 times the mapping speed. The scientific motivations and goals for BLAST, outlined below, are similar to those of FIRST and are achievable within 3–5 years with a series of LDB flights:

- Conduct large-area extragalactic 250–500 μm surveys and detect ~ 150 and ~ 1500 high-z galaxies in the test flight (6 hours) and long-duration flight surveys (50 hours) respectively (see Table 1);

- Measure the confusion noise at 250–500 μm thereby laying the foundation for future BLAST and FIRST survey strategies. It will also allow a study of the clustering of dust-emitting galaxies over an important range of angular scales;

- Combine the BLAST 250–500 μm spectral energy distributions (SEDs) and source counts with those of SCUBA at 850 μm and MAMBO at 1.3 mm. This will determine the redshifts, rest-frame luminosities, star formation rates (SFRs), and evolutionary history of starburst galaxies in the high-z universe. It will also identify the galaxy populations responsible for producing the far-IR background;

- Conduct Galactic surveys of molecular clouds and identify dense, cold pre-stellar (Class–0) cores associated with the earliest stages of star formation (see Fig.2). For example in < 6 hours BLAST can conduct a $\sim 50\,\mathrm{deg}^2$ 250 μm survey of molecular clouds with a 3σ sensitivity of

Table 1. Estimated number of galaxies detected in test flight and LDB surveys with primary apertures, D, of 2.0 m and 2.5 m respectively. Illustrative redshift distributions for galaxies detected with a S/N > 5 are included. The entire LDB flight will be 250 hours allowing for several such surveys.

6 hour 300 μm test-flight survey strategies: D=2.0 m, $\theta = 41''$					
survey area (sq. degrees)	1σ depth	no. of detected galaxies		no. of > 5σ galaxies	
		> 5σ	> 10σ	$z > 1$	$z > 3$
0.24	7 mJy	120	34	110	18
0.55	10 mJy	150	40	135	20
1.1	15 mJy	135	30	125	16
4.4	30 mJy	120	30	110	13

50 hour 300 μm LDB survey strategies: D=2.5 m, $\theta = 32''$					
survey area (sq. degrees)	1σ depth	no. of detected galaxies		no. of > 5σ galaxies	
		> 5σ	> 10σ	$z > 1$	$z > 3$
1.7	5 mJy	1420	450	1300	250
3.3	7 mJy	1670	480	1530	250
6.8	10 mJy	1870	500	1680	250
15.4	15 mJy	1890	420	1740	220
61.6	30 mJy	1680	420	1530	180

300 mJy, equivalent to a mass sensitivity $\sim 0.05\,\mathrm{M_\odot}$. Combining the 250-500 μm BLAST data with SCUBA data at 850 μm will determine the radial density and temperature structures of the pre-stellar cores which are sensitive to the details of the clouds collapse[17];

- Survey the diffuse Galactic emission and make detailed comparisons with surveys at longer and shorter wavelengths (CGPS, SCUBÁ, MSX);

- Observe objects within our solar system including the planets and large asteroids.

3 Long Duration Balloon (LDB) Surveys

The 250 hour LDB flights will conduct a complementary series of extragalactic and Galactic surveys of varying depths and areas. The increased observation

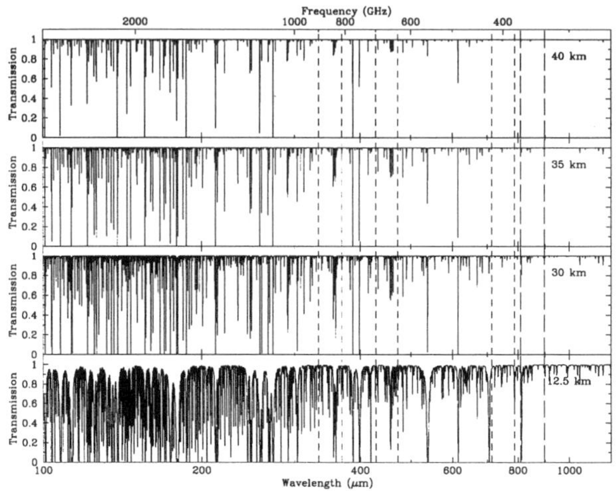

Figure 1. Atmospheric transmission at balloon and SOFIA altitudes.

time, compared to the test-flight, and the planned addition of a larger 2.5 m primary will increase sensitivities and resolution while lowering confusion limits, thereby improving the statistics of the brighter sub-mm detections.

The uniqueness of BLAST is that it has sufficient sensitivity to measure the peak of the redshifted SED of luminous starburst galaxies at $z = 1 - 5$, at shorter wavelengths (250-500 μm) than existing imstrumention. Unlike SCUBA or MAMBO observations, BLAST is capable of directly constraining the rest-frame FIR luminosity and, with some assumptions, the star formation rate. The combination of BLAST, SCUBA and MAMBO surveys, complemented by deep radio (VLA, GBT) and X-ray (AXAF,XMM) will significantly improve the ability to discriminate between the various evolutionary models.

As Table 1 indicates, the improved BLAST instrument will detect thousands of high-z galaxies during one of several 50 hour surveys in a single LDB flight. Possible LDB extragalactic survey targets include: (i) for Northern hemisphere flights, the Lockman Hole, and ELAIS N2, ELAIS N1, the HDF-North and flanking fields and future LMT 1100 μm surveys; (ii) for Southern hemisphere flights, the MARANO field and the HDF-South.

We will also survey a large ($> 25\%$) fraction of the Galactic plane concentrating on regions undergoing high and low-mass star formation.

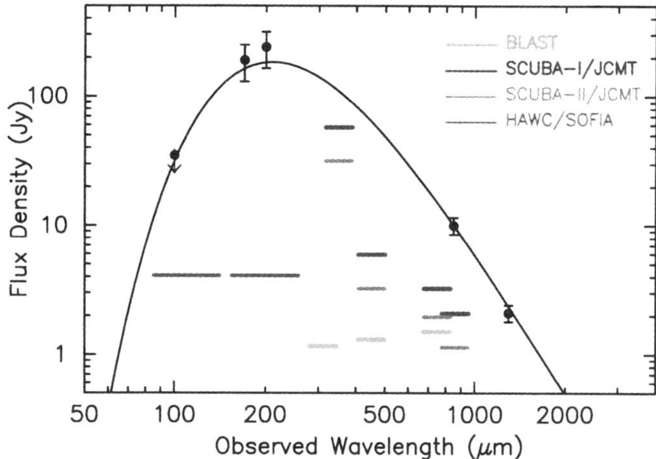

Figure 2. The spectral energy distribution of L1544, a pre-stellar embedded core. The horizontal bars represent the $10\,\sigma$ 1 sec sensitivities to extended emission.

4 Instrument

The design and specifications of BLAST are driven by the science goals, availability of existing instrumentation and the practical limitations of balloon flights. The following sections describe the instrumental requirements needed to meet the science goals.

The decision to use an LDB platform for this experiment takes into consideration a combination of sensitivity, cost, and time-scale. When comparing this experiment to ground-based or airborne observations, the clear advantage is greater atmospheric transmission at balloon altitudes (35-40 km, Fig.1). The high atmospheric emission at lower altitudes (SOFIA and SCUBA) limits the instrument sensitivity and, in some of the higher frequency bands, makes the measurement virtually impossible without the use of sub-orbital and orbital platforms.

4.1 Optical Design

The BLAST gondola is designed to hold a 2.5 m diameter mirror. The test flight will use a smaller 2.0 m spherical mirror currently being fabricated as part of the FIRST development. The secondary mirror will be designed to give diffraction limited performance over a $10' \times 10'$ FOV at the Cassegrain focus

Table 2. Telescope and receiver parameters

Telescope:	Temperature	300 –230 K
	Used diameter	1.9 m
	Emissivity	0.04
Detectors:	Bolometer optical NEP	3.0×10^{-19} W Hz$^{-1/2}$
	Bolometer quantum efficiency	0.8
	Bolometer feed-horn efficiency	0.7
	Throughput for each pixel	$A\Omega = \lambda^2$ (2fλ feed-horns)
Bolometers:	Central wavelengths	250 350 500 μm
	Number of pixels	149 88 43
	Beam FWHM	30 41 59 arcsec
	Field of view for each array	6.5×13 arcmin
	Overall instrument transmission	30%
	Filter widths ($\lambda/\Delta\lambda$)	3
	Observing efficiency	90%

at $\lambda = 300\,\mu$m. The estimated antenna efficiency is $\geq 80\%$ and is determined by a combination of the $10\,\mu$m rms surface roughness of the primary and the quality of the re-imaging optics.

Radiation from the telescope will enter the cryostat through a 5–6 cm diameter vacuum window near the Cassegrain focus. We will use a telescope focal ratio of f/5 to place the position of the focus $\simeq 20$ cm behind the central hole in the primary. The window will be made from 50 μm thick polypropylene that has $< 0.1\%$ loss. Blocking filters at the intermediate cold stages of 77 K and 20 K will reduce the radiation loading on the LHe to < 10 mW.

Table 3. BLAST loading, BLIP noise and sensitivities

Band (μm)	250	350	500
Backgrnd. power (pW)	25.6	18.3	13.5
Backgrnd.-lim. NEP (W Hz$^{-1/2}$)	20	14	10
NEFD (mJy s$^{1/2}$)	236	241	239
ΔS(mJy) 1σ 1hr, 1deg^2	38	36	36
ΔS(mJy) 1σ 6hr, 1deg^2	15.5	14.7	14.6

The radiation will be re-imaged onto the detector arrays using a pair of cooled off-axis parabolic mirrors arranged in a "Gaussian beam telescope" configuration. This configuration makes additional corrections for aberrations in the main telescope and provides a flat focal plane with phase centers independent of wavelength for single-mode Gaussian beams. A cold aperture or Lyot stop will be located between the two re-imaging mirrors at the position of an image of the primary mirror to provide additional sidelobe rejection. The second parabolic mirror has a focal length equal to its distance from this aperture to insure that all of the detectors in the array have illumination centered on the Lyot stop. The off-axis angle of the final re-imaging mirror allows us to place dichroic beam splitters in front of the focal plane making possible simultaneous measurements by different arrays at different wavelengths. For the first flight of BLAST, we plan to use detector arrays similar to those being developed for the SPIRE [8] instrument on FIRST.

4.2 Detectors.

The BLAST focal-plane will consist of arrays of 149, 88 and 43 detectors at 250, 350, and 500 μm respectively. The detectors will be silicon nitride micromesh ("spider-web") bolometric detectors coupled with $2f\lambda$ feedhorn arrays.[4] The detector and feedhorn technology is well established and has been tested using the BOLOCAM instrument in late 1999. The sensitivity of the detectors is limited by photon shot noise from the telescope and atmospheric emission. Table 2 gives the telescope and detector parameters. We estimate a total emissivity for the warm optics of $\simeq 5\%$ dominated by blockage from the secondary mirror and supports. We estimate the optical efficiency of the cold filters and optics to be $\epsilon_{\mathrm{opt}} \geq 0.2$ based on measurements with similar bolometers coupled to horn arrays at millimeter wavelengths in the BOLOCAM test dewar. The estimated detector NEFD's assuming a 10% bandwidth are given in Table 3

5 Conclusion

BLAST will be the first balloon-borne instrument to take advantage of a new generation of bolometric arrays. The transparency of the atmosphere at balloon altitudes will allow it to observe in the sub-millimeter band where measurements are difficult from the ground. The combination of state-of-the-art detectors on a sub-orbital observing platform will give us the opportunity to collect data which will revolutionize our view of the sub-millimeter sky. This class of intermediate missions will provide an integral step towards our understanding of the evolution of stars, galaxies and clusters in the pre–FIRST era. After an initial test flight in 2002, the instrument will have its first Long Duration Balloon flight from Antarctica in 2003.

References

1. Andre, P., Ward-Thompson, D.W., Barsony, M., *ApJ* **406**, 122 (1993).
2. Bachiller R., Tafalla M., *The Origin of Stars and Planetary Systems*, eds. C.J. Lada, N.D. Kylafis, (Kluwer), 227 (1999).
3. Beichman, C.A., Myers, P.C., Emerson, J.P., *ApJ* **307**, 337 (1986)
4. Bock, J.J., Glenn, J., Grannan, S.M., Irwin, K.D., Lange, A.E., Leduc, H.G., Turner, A.D., *Proc. SPIE* **3357**, 297 (1998).
5. Bock, J.J., Leduc, H.G., *The Far Infrared and Submillimetre Universe*, ed. Wilson, A., 349 (1997).
6. Chandler C.J., Richer J.S., *ApJ* **530**, 851 (2000).
7. Chapman, S.C., Scott, D., Borys, C., Fahlman, G.G., *MNRAS*, astro-ph/0009067, submitted (2000).
8. Griffin M., Swinyard B., Vigroux L., in Munich SPIE Symposium, in press (2000).
9. http://www.sofia.usra.edu/observatory/instruments/first_light/hawc_abstract.2.html
10. Holland, W.S., Robson, E.I., Gear W.K., Cunningham, C.R., Lighfoot, J.F., Jenness, T., Ivison, R.J., Stevens, J.A., Ade, P.A.R., Griffin, M.J., Duncan, W.D., Murphy, J.A., Naylor, D.A., *MNRAS* **303**, 659 (1999).
11. Hughes, D., Serjeant, S., Dunlop, J., Rowan-Robinson, M., Blain, A., Mann, R.G., Ivison, R., Peacock, J., Efstathiou, A., Gear, W., Oliver, S., Lawrence, A., Longair, M., Goldschmidt, P., Jenness, T., *Nature* **394**, 241 (1998).
12. Hunter, T.R., Benford, D.J., Serabyn, E., *PASP* **108**, 104 (1996).
13. Lamarre, J.M., Giard, M., Pointecouteau, E., Bernard, J.P., Serra, G., Pajot, F., Desert, F.X., Ristorcelli, I., Torre, J.P., Church, S., Coron, N., Puget, J.L., Bock J.J., *ApJ* **507**, L5 (1998).
14. Lilly, S., Eales, S., Gear, W.K., Hammer, F., Le Fevre, O., Crampton, D., Bond, R., Dunne, L., astro-ph/9901047, in press (1999).
15. Mardones, D., Myers, P.C., Tafalla, M., Wilner, D.J., Bachiller, R., Garay, G., *ApJ* **489**, 719 (1997).
16. Smail, I., Ivison, Blain, A.W., *MNRAS* **490**, L5 (1997).
17. Ward-Thompson, D.W., Scott, P.F., Hills, R.E., Andre, P., *MNRAS* **305**, 143 (1999).

SCUBA–2, THE NEXT GENERATION, WIDE–FIELD SUBMILLIMETRE CAMERA FOR THE JCMT

IAN ROBSON AND WAYNE HOLLAND

Joint Astronomy Centre,660 N. A'Ohoku Pl., Hilo, HI 96720, USA
E–mail:eir@jach.hawaii.edu, wsh@jach.hawaii.edu

WILLIAM DUNCAN

UKATC, Royal Observatory, Blackford Hill, Edinburgh, Eh9 3HJ, UK
E–mail: william.duncan@roe.ac.uk

Following on from the enormous success of SCUBA on the JCMT in totally revolutionizing submillimetre continuum astronomy, its successor, SCUBA–2 is now well into the development phase. SCUBA–2 will be a simultaneous dual–waveband, wide–field imager, having an unvignetted field–of–view of 64 arcminutes square, limited in sensitivity by the sky background alone, reaching the confusion limit in around an hour. SCUBA–2 will utilize an array of superconducting TES devices built by NIST and the University of Edinburgh. The array will be directly illuminated and the pixels will be half the diffraction spot diameter, requiring 25,600 pixels at 450 microns and 6,400 at 850 microns. This architecture will allow full diffraction resolution imaging without the need to jiggle the secondary mirror. SCUBA–2 passed its Conceptual Design Review in 1999 and the detector architecture downselect in May 2000, and has been enthusiastically supported by the JCMT user community, the JCMT Board
and PPARC. The project is led by the prime contractors the UKATC, and, like SCUBA, involves close collaboration with QMW.

1 Introduction

SCUBA [1] is undoubtedly one of the most important instruments ever built for astronomy. As well as being the first large–scale array for submillimetre astronomy, it is a true facility instrument. SCUBA is mounted on the world's largest submillimetre telescope, which is well supported by scientific and technical staff; it has a dedicated suite of data reduction software making data analysis and hence publication of the results readily achievable. As such, SCUBA has opened up the submillimetre, perhaps the last unexplored window on the Universe. Three areas of science stand out as truly revolutionary: galaxy evolution in the early Universe [e.g. 2,3,4]; dust disks around main–sequence stars [e.g. 5,6]; large–scale survey programmes addressing star formation [7,8].

In 1998 the JCMT Board set up an International Review to address the future scientific direction and competitiveness of the facility. In preparation for that review, one of us (IR) looked at the competition SCUBA would be facing from other facilities and also the science that would be needed to be done. One thing was clear, a larger and faster array was the obvious next development. Although SCUBA continues to be upgraded to improve its performance and reliability, there are natural limitations beyond which it is more cost effective to build a new instrument rather than continue upgrading. Again, one of us (IR) issued a challenge to see if it was possible to construct a much larger array: one that would fill the field–of–view of the JCMT; that would have no moving parts (and hence would be simpler to build and operate than SCUBA); would operate at helium–three temperatures (as opposed to 100mK for SCUBA); would use state–of–the–art (purchased) arrays and would be, in effect, the first CCD–type of submillimetre imager, potentially allowing stare–mode of observing without the AC chopping/nodding of previous devices. Two of us (WH & WD) responded to the challenge and came up with a pre–proposal plan which eventually developed into what will be presented here as SCUBA–2.

2 Why we need a SCUBA–2

Many key scientific programs require large areas to be imaged in order to provide unbiased surveys of sources and source structure. SCUBA has a limited field–of–view (fov) of only 2.3 arcminutes, which each of the two arrays simultaneously samples. This immediately poses limitations on the key science programs that can be undertaken in reasonable times, even on the JCMT with its fully flexible scheduling mode of operation and on the excellent site of Mauna Kea. Thus, the main reason for a new SCUBA is the requirement for a much larger field–of–view. The unvignetted fov of the JCMT at Nasmyth focus is a circle of diameter just over 11 arcminutes. Although the optical design challenges are severe, it is a key goal that a wide–field replacement for SCUBA to use a minimum of an 8 by 8 arcminute patch of the unvignetted focal plane. This is the maximum size of square array that will fit the field of view without having obscured pixels. Furthermore, the throughput of SCUBA is less than the design goal and so there is scope for improving the throughput and hence raw pixel sensitivity, so that a SCUBA–2 would be truly photon–noise limited by the background sky. As the JCMT moves into a new era in which it will operate as part of the Smithsonian Submillimetre Array (to give subarcsecond imaging) and has a suite of heterodyne cameras, the time available for continuum imaging will be reduced. Hence, being able to map and image faster is a clear goal for facility productivity. Such a device would also keep the JCMT at the forefront of submillimetre astronomy, a key factor for the funding agencies of the UK, Canada and the Netherlands.

However, achieving such an ambitious goal is only possible if the arrays can be procured, and this was the key factor in the early phase of the project. The fact that such arrays have now been demonstrated to work in the lab, and can be constructed is the breakthrough that changes SCUBA–2 from an interesting design study to a feasible instrument.

3 The Scientific Goals

The user community was polled with respect to their interest in a SCUBA–2 opportunity and the response was extremely encouraging, in fact amazingly so. A huge range of science goals were identified, from solar system to cosmology, and these are listed on the scientific case for SCUBA–2 which is presented on the SCUBA–2 web–pages of the JCMT. In terms of operational goals, the key requirements are: the ability to reach the confusion limit in an hour or two enabling very deep imaging to be carried out; the ability to map, to a reasonable depth, an area of several square degrees in a few hours enabling large–scale mapping to be undertaken; to have dual waveband capability for spectral index determination of dust properties (and in the context of this meeting – redshifts). Finally, SCUBA–2 must be sensitive to point–source photometry of known objects enabling the spectral energy distributions to be determined. This latter goal implies that mapping speed cannot be traded for a loss of pixel sensitivity.

The goals were then refined to a specification of a mapping speed of at least 100 times that of the current (*upgraded*) SCUBA, a pixel sensitivity at least 50% better than the *upgraded* SCUBA and simultaneous operation at two wavelengths. DC operation was preferred with no sky–chopping. High image fidelity and map dynamic range were also important goals, all of which can be summarized by "faster, deeper, better". This will allow SCUBA–2 to undertake very deep, but large–scale extragalactic surveys (crucial for the topic of this meeting) and very large–scale, high–resolution Galactic surveys. Table 1 gives an indication of examples of SCUBA–2 performance assuming the baseline design parameters.

Table 1. Examples of SCUBA–2 performance

Example (λ=850μm)	Integration time (hours)	
	SCUBA	SCUBA–2
Point–source photometry to 5–σ flux limit of 2 mJy	7.3	0.6
Map of the Hubble Deep Field (N) to noise of 0.5 mJy	32	0.5
Galactic plane survey (20×2°) to noise level of 30 mJy	850	0.9
Survey 5° diameter molecular cloud to noise of 10mJy	4700	5
Deep extragalactic survey of a 1 deg² area to noise level of 0.5mJy	22,000	23

4 Baseline specification

SCUBA–2 will consist of two arrays operating simultaneously at currently preferred wavelengths of 850 and 450 microns and they will be diffraction limited at each wavelength. The field–of–view will be a minimum of 8x8 arcminutes, the detectors will be background limited from the sky and the pixels will be DC coupled. This field–of–view, of some 64 arcminutes^2, compares extremely well with other facilities

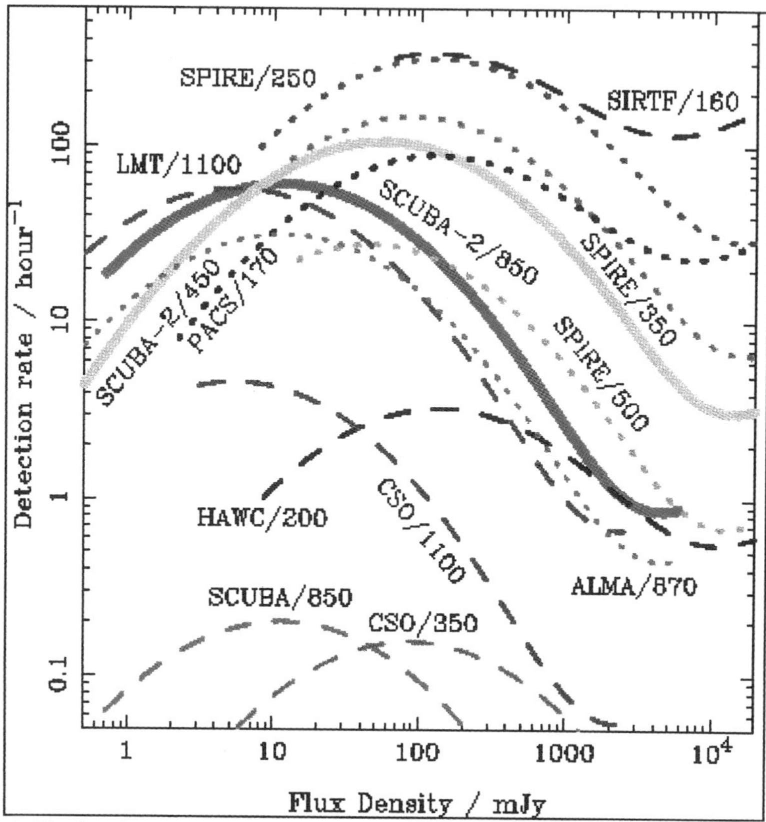

Figure 1. The detection rates as a function of 5σ depth for a variety of instruments. The lines stop at the left–hand side due to confusion limits, and on the right where there is only one source in the sky. Note the highly competitive nature of SCUBA-2 for the deepest surveys, where the smaller spaceborne telescopes lose out to confusion. Diagram courtesy of Andrew Blain (see later this volume).

and coupled with the pixel sensitivity means that SCUBA-2 is very competitive with any instrument currently proposed (see Fig. 1). Indeed, SCUBA-2 also turns out to be extremely complementary to many other planned facilities, specifically the space and airborne observatories and in addition, ALMA. This latter point is particularly important as although ALMA reaches a very deep level, and has a much lower confusion limit, its instantaneous field–of–view is extremely small and so having deep surveys to act as finding sources for ALMA follow–up is a key role for SCUBA-2. This is also the case for FIRST, albeit for different reasons.

Not to be forgotten the importance of imaging polarimetry has not been neglected and this will be possible with SCUBA-2, in the same manner as currently undertaken by SCUBA.

5   Instrument design

5.1   Detectors

The heart of SCUBA-2 is the detector arrays. These will use the next generation detector technology to enable large format arrays to be realized. The detecting element will be a transition edge sensor (TES); these will be voltage biased and superconducting. The key point about a superconducting TES is that in the transition region between the superconducting and normal states, the resistance is an extremely steep function of the temperature. Hence such a device, coupled to an antenna to absorb photons is an extremely sensitive temperature sensor. Any change in the temperature (due to absorption of photons) of a TES film held at a constant voltage bias, results in a change of current which can be amplified and measured. Another key property of voltage biased TESs is that they are self-biasing. This is extremely important in ensuring that an array can be biased using a single bias supply despite small variations in the critical temperatures of the superconducting-normal bi-layers of each device during manufacture on the wafers. Self-biasing is achieved because the bias power is a source of heating for the device and as the temperature of the film is reduced below the critical temperature, the resistance falls sharply and the joule heating increases. This is a negative feedback loop in which an initial increase in temperature and hence resistance due to an absorption of photons is rapidly opposed by a decrease in the bias current and the Joule heating and a return to the original device temperature. Equilibrium is rapidly achieved when the heating and cooling are matched – these devices are true power meters. The extremely sharp transition region also means that this electro-thermal feedback effect can significantly speed up the device compared with the physical time constant for thermal relaxation not assisted by the feedback effect.

Transition edge sensors do not come without risk and complications, however. TESs require magnetic shielding, there is a question over their 1/f performance (which might limit the DC coupling requirement), there is a question about how robust they will be under repeated thermal cycling to helium temperatures, and finally, there is the question about their dynamic range when exposed to an uncertain power loading from photon illumination.

In all large array devices for the submillimetre, where the detectors are at helium temperatures, a key practical issue is the electrical leadouts from the array to the outside of the cryostat. Indeed, this is the limiting factor the size of arrays unless multiplexing is used. Multiplexing is an integral part of the SCUBA-2 novelty, and amplification and multiplexing is enabled through SQUIDS [9]. The detector device architecture that was eventually selected is similar to that which has been in development for the SPIRE detectors for the FIRST space mission [10] [11]. The construction of the arrays is very complex and is not covered here suffice to say that the array is in two parts. The top part is the micro–machined bolometer array, this is bump–bonded with indium to the lower SQUID multiplexer and interconnect chip. Needless to say, producing such a challenging, bonded component was one of the key factors in determining the detector architecture (see next section). The detector resides at 100mK, requiring additional technology such as a dilution refrigerator. Ribbon cables take the signals to the next stage at 0.8K where SQUID arrays perform amplification before the final amplification stage at 300K.

5.2 *Detector architecture*

The above description refers to the fully–filled, 0.5Fλ architecture that came out of the detector downselect review on May 4th 2000. The basic choices for a large format array boil down to either the traditional format of 2 Fλ spaced pixels fed by feedhorns (such as SCUBA), or, a filled array with the pixels spaced by half the diameter of the diffraction disk size (0.5 Fλ) and illuminated by direct radiation (as in a CCD for example). There are pros and cons to using each of these along with attendant risks, and so a detailed risk analysis and cost–benefit analysis was undertaken leading up to the downselect. It is immediately obvious that the filled array is more susceptible to stray light and so great care must be taken to define the illumination from the telescope by a (very) cold stop. In turn, this brings the benefit of producing a top–hat illumination of the telescope by the detector. Further disadvantages of a filled array are that it has a lower speed for detecting known point sources, it has greater vulnerability to stray radio–frequency signals, 12 times more detectors are required for a given field–of–view, and the photon noise–equivalent–power for each detector must be lower by a factor of around two because of the reduced background power per pixel. On the other hand, the advantages are large: it produces a much higher speed for mapping; it provides instantaneous sampling of the image without jiggling of the telescope secondary mirror; it has a slightly narrower beam on the sky due to the illumination function.

The advantages were agreed to outweigh the disadvantages and as the risks were not dissimilar for both architectures the bare–fully–filled–array was selected. As noted above, as well as the implications of operate temperature and bump–bonding, the number of pixels needed is immense: 6,400 at 850 microns and 25,600 at 450 microns. The arrays will be constructed in sub–arrays and butted together (with no gaps) in the focal plane.

5.3   Optical arrangement

SCUBA–2 will be fed by the tertiary mirror of the JCMT, through a set of relay optics to the Nasmyth platform. In fact the optics for SCUBA–2 are far from trivial due to the aberrations in the focal plane of the telescope, the large fov and the fact that in the simplest mechanical and cryogenic arrangement, the optics on the antenna are optically sheared as the telescope elevation changes during observing To make calibration and re–gridding data from Nasmyth coordinates to RA/DEC easy, good image quality and low field distortion are required. Key aspects of the internal system of optics will be the need for high quality baffling and control of stray light as noted above.

6   Status

The Conceptual Design Review was held in September 1999 and the outcome was presented to the JCMT Advisory Panel and Board in November, where it was enthusiastically received and endorsed as the highest priority scientific project for the JCMT. Funds were provided for the detector architecture selection phase, which was completed on May 4th 2000. In the meantime, the UK Particle Physics and Astronomy Research Council (PPARC) approved proof–of–concept funding for a prototype array. The JCMT Board will return to address the funding of SCUBA–2 at the forward look meeting in November.

The prime contractor for the construction of SCUBA–2 is the United Kingdom Astronomical Technology Centre (UKATC) at the Royal Observatory Edinburgh, with collaboration from Queen Mary Westfield College (London). The detectors will be provided under contract by the National Institute of Standards and Technology (NIST) at Boulder, Colorado in collaboration with the University of Edinburgh Micromachining Laboratory. The project scientists has just been appointed, Dr Wayne Holland (JCMT) and the project is now underway. Delivery of this exciting and world–beating instrument is expected to the JCMT in late 2005.

References

1.   Holland,W.S. et al..   SCUBA: a common–user submillimetre camera operating
     on the James Clerk Maxwell Telescope. MNRAS, **303** (1999) pp. 659–672.

2.   Smail,I., Ivison,R.J. and Blain,A.W. A Deep Submillimeter Survey of Lensing

     Clusters: A New Window on Galaxy Formation and Evolution. Ap.J. **490**, (1997) pp. L5–L8

3. Hughes,D.H. et al. High–redshift star formation in the Hubble Deep Field
     revealed by a submillimetre wavelength survey. Nature **394** (1998) pp241–
247.

4. Barger,A.J. et al. Submillimetre–wavelength detection of dusty star–forming
    galaxies at high redshift. Nature **394** (1998) pp. 248–251.

5. Holland,W.S. et al. Submillimetre images of dusty debris around nearby stars.
    Nature **392** (1998) pp. 788–791.

6. Greaves,J. et al. A Dust Ring around ε Eridani: Analog to the Young Solar
    System. Ap.J. **506** (1998) pp. L133–L138.

7. Johnston,D. and Bally,J. Submillimeter Wavelength Imaging of the Integral–
    shaped Filament in Orion. Ap.J. **510** (1999) pp. L49–L54.

8. Pierce–Price,D. et al. A SCUBA submillimetre survey of the Galactic Centre
    ApJ., (2000) submitted.

9. Chervenak,J.A. et al., Superconductor multiplexer for arrays of transition
edge
     sensors. Appl.Phys.Lett. **74** (1999), 26–

10. Griffin,M.J. et al. Spire–a bolometer instrument for FIRST. Proc. SPIE 3357
     (T.Phillips, ed.) (1998) pp. 404–413.

11. Bock,J.J. et al. Silicon nitride micromesh bolometer arrays for SPIRE. *Ibid.*
     pp. 297–304.

SEMI RIGID AND ADJUSTABLE CFRP MEMBRANE FOR RADIO TELESCOPE APPLICATIONS

D. GILES AND S. KULICK

Composite Optics, Inc.
9617 Distribution Ave. San Diego, CA 92121
NSF SBIR Phase II Contract DMI–9801169

"This material is based upon work supported by the National Science Foundation under Award Number: DMI–9801169. Any opinions, findings, and conclusions or recommendations expressed in this publication are those of the authors and do not necessarily reflect the views of the National Science Foundation."

1 Abstract

High accuracy, very lightweight carbon fiber reinforced plastic (CFRP) membrane reflectors have been demonstrated by Composite Optics, Inc. (COI) for applications in ground–based, millimeter wave radio astronomy. The figure of a semi–rigid membrane can be significantly improved by suitably altering the adjustment of an array of passive adjusters that support the membrane. A factor of ten contour improvement is possible. First, the surface is measured and analyzed for Zernike error modes. Then an adjustment scenario, dictated by a customized software program, is applied to the adjusters. This process is repeated to produce dimensionally stable reflectors several meters in diameter at or below 0.001 inch rms.

Keywords: adjustable reflector, CFRP, millimeter wave radio astronomy

Millimeter wave radio astronomy is of great interest to the scientific community and particularly to the National Radio Astronomy Observatory (NRAO). NRAO currently has plans to produce a radio telescope to operate between the frequencies of 30 and 900 Ghz. The radio telescope, known as the Millimeter Array (MMA), will consist of an array of 36 ten meter diameter parabolic reflector surfaces and will be located at Llano de Chajnantor, a plateau at 16,400 feet elevation, in Chile. This shorter wavelength instrument places a significant burden on the surface accuracy and thermal performance of the antenna.

Attempts have been made to construct the antenna reflective surfaces from carbon fiber composite panels in a sandwich construction using aluminum honeycomb as the core. In one notable example in France, each panel, in the order of a meter square, was supported at each of four corners and adjusted in tip, tilt and focus to form an accurate (25μm rms) reflecting surface. The panels themselves were relatively rigid structures. Unfortunately, this approach, although lightweight, reasonably stable over temperatures and sufficiently accurate, was not without problems. The reflective surface was a very thin layer of vapor–deposited aluminum (VDA) on the composite surface. The installation, based on technology several years old, had suffered from some erosion and corrosion problems. In spite of the presence of a protective coating, the surface was punctured in several locations. Each of these openings became a corrosion site. There have also been reports of delamination between the carbon fiber–reinforced plastic (CFRP) composite faceskins and the aluminum core. With this as anecdotal background, it is not surprising that there are questions about future use of a CFRP as an integral part of the reflecting surface. These concerns do have merit. It is also true that there have been recent advances on several fronts that justify examining the use of CFRP for the reflector membrane.

Composite Optics, Inc. (COI) has developed a novel and low–cost method for fabricating large, high–accuracy reflector surfaces. Dubbed the adjustable membrane approach, this method entails fabricating a thin and uniform membrane surface and later adjusting the contour with an array of adjusters to within acceptable limits. Since the panels are adjusted as a final operation, the inherent difficulties and expenses associated with fabricating metal–coated rigid panels are eliminated. Analytical results and empirical data both substantiate achievable contour improvements of an order of magnitude or more. Thus an adjustable panel for the MMA radio telescope may be manufactured at 50 to 127μm rms and adjusted within tolerance to 12μm rms.

COI's adjustable membrane concept could replace the rigid panels and not suffer from the inherent durability issues connected with aluminum honeycomb core sandwich construction and thin metallic surface coatings. Finally, alternate metallization processes not possible with a rigid body panel will use a much thicker metal layer and will produce a more durable reflective surface.

2 Approach

The feasibility of producing rigid, composite, sandwich panels for high accuracy reflector surfaces has been demonstrated by MAN Technologies of Munich (Reference 1), Germany and Mitsubishi Electric Corp. of Tokyo, Japan (reference 2), and by COI. The costs of manufacturing highly accurate reflector surfaces may be significantly reduced if this new approach is taken.

A solid CFRP laminate has significant in–plane stiffness but is able to deflect under small transverse loads. Therefore, a reflector panel having a certain error as manufactured can be distorted by an array of passive adjusters to improve the surface figure. In this concept, a reflector surface is manufactured from conventional pre–impregnated (prepreg) carbon fiber reinforced epoxy material by a hand layup process and cured in an oven or autoclave. The reflector membrane is a uniform, isotropic layup of CFRP and does not use any honeycomb core material. Because the manufacturing error will be adjusted out at a later time, the issues of mold thermal expansion and thermal lag during cure are of small significance as are the effects of internal stresses caused by temperature gradients upon heat up and cool down. A thin layer of aluminum foil, between 0.002 and 0.005 inch thick, is applied to the CFRP membrane on both the front and back surface and the part is molded in such a way that the metal surface exactly matches the figure of the layup mold. The foil creates an electrically conductive layer for microwave reflectivity that is significantly thicker and more durable than the VDA or aluminized plastic film coating typically used in a rigid panel concept. The foil method works in this capacity because the membrane surface will be adjusted after the manufacturing process; otherwise, the effect of dissimilar thermal expansion between the metal and composite would yield unacceptable contour results in a rigid panel concept.

To verify that a reflector surface may be adjusted from the as–manufactured figure to the requirement of 12 μm rms, an analytical model was developed using MSC Nastran finite element modeling (FEM) software. A model of the parabolic reflector surface was created and a series of unit loads was applied to each of the possible adjuster node locations in the model. The axis symmetry of a center–fed paraboloid was employed to minimize the number of FEM load cases required, thus a single radial set of applied loads was extrapolated to characterize the influence of a unit load applied at any adjuster location for the entire surface of the membrane. The resultant influence coefficient matrix was used to calculate the required load application at each adjuster node; this minimizes the total surface rms error for a given set of adjuster coordinates and a database of measured (or assumed) surface figure errors. Countless model cases and analyses led to the following conclusions regarding the adjustability of a 3.3M reflector:

- At 0.12 inches thick, a large number of adjusters (> 258) will be needed to bring an initial 127 μm rms (assumed) into acceptable limits (< 12 μm).

- Thicker membranes are "better" from an adjustment standpoint.

- The 0.20 inch thickness shows potential for adjusting while utilizing a smaller amount of composite.

- Adjuster placement can be manipulated to specifically reduce certain modes.

- Specific rms targets can be used to define the minimum number of adjusters needed.

Dr. John Kibler of COI has written an integrated, Windows based program to perform all of the calculations and analysis required to inspect, analyze, diagnose, and adjust a reflector surface. The software also includes many other engineering and analysis capabilities commonly used by COI reflector engineering. Since the adjustment analysis was added to an already existing program, its capabilities inherently address the unique requirements for building and analyzing Zernike shaped reflectors, as well as spheres, parabolas, hyperbolas, planes, and cylinders.

Given that a reflector surface is required to be iteratively inspected and adjusted, and given the measurement precision required to verify a large reflector to 12 μm (0.0005 inch) rms, COI investigated an alternative method for inspecting the reflector surface. The baseline inspection device used at COI for large reflectors is a multi–headed theodolite system; however, a laser tracker device was brought in for evaluation and use in the fabrication of the test–bed adjustable reflector. A laser tracker is a single–headed device used to digitize a surface or component feature dimensions. It is similar to a single theodolite head in that it uses a pair of shaft encoders to measure elevation and azimuth angles. Unlike a theodolite, which uses two plus heads and triangulation to solve for a component dimension, a laser tracker uses a laser interferometer to measure distance. Thus with two angles and a distance measurement, a single head instrument can generate coordinates in three space. In operation, the laser beam is reflected back to the instrument by a spherically mounted retro–reflector (SMR). As the SMR is moved, the laser beam is steered by elevation and azimuth motors; the system tracks the SMR by maintaining the maximum reflected beam intensity. The SMR is dragged across a surface or feature while the tracker records the (x,y,z) coordinates at the center of the sphere over time.

COI used a Spatial Metrix Corp. (SMX) laser tracker, model 4500, which was also used for inspection and adjustment of the technology demonstrator reflector. A typical inspection map consisted of 6,000 to 8,000 points. Data points were collected with a .5 inch distance spacing in the x–direction and about a 2 to 4 inch spacing in the Y direction of the reflector. Figure 1 is a graphical representation of the surface error of the test bed reflector, including a '+' at each measurement point.

The laser tracker device performed exceptionally well, both in reducing the time required to perform an inspection and in improving the measurement accuracy capability over a theodolite system. Each surface inspection required one operator and took only 20 minutes from start to finish. This represents nearly a 60X improvement in data collection rate. The SMX laser tracker system and software offered some additional capability that was also taken advantage of during the inspection and adjustment operations performed on the breadboard reflector.

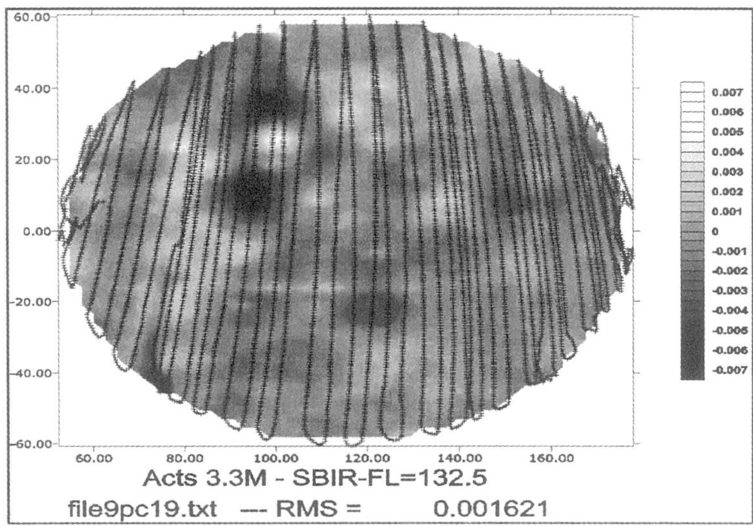

Figure 1. Inspection map and measured points.

3 Results

Following the adjustment analysis and the development of an integrated inspection and adjustment software program, COI manufactured the test bed reflector to the following specifications:

Surface[1]:

- Shape: Parabolic surface of revolution

- Focal length = 132.5 inch

- Aperture = 3.3M Ø, offset

- Offset = 50 inch +Y direction

Laminate:

- Standard modulus (33 Msi) carbon fiber

- Unidirectional prepreg tape

- 0.200 inch thick, 16 plies, quasi–isotropic

- Between 57 and 107 adjusters

The reflector described above was manufactured at COI using low–cost CFRP materials and processing methods developed specifically for application in terrestrial, high precision reflector systems. After manufacture, it was supported on a rigid structure with passive micro–adjusters. The SMX laser tracker was used to inspect the surface and the integrated software program was used to assist with the adjustment analysis and prescription for the displacement and number of turns at each micro–adjuster.

While the mold tool had a 0.0043 inch rms with a 132.5 inch focal length, the first inspection of the reflector assembly (after trimming the edge of the laminate to size) revealed a 0.0078 inch rms. The increase in surface figure distortion is an inherent function of the mechanical and physical properties of the lower cost materials and processes selected for this application; to reiterate the direction is not to produce a rigid body reflector panel that comes off the mold immediately in tolerance. Beginning with only 57 adjusters on a 16–inch square grid (0.56 adjusters per square foot) the surface was adjusted to 0.0019 inch rms. Though the analysis predicted a 0.0016–inch rms was achievable given the quantity and position of the adjusters, more adjusters were added to gain better control of the reflector contour around the perimeter and to approach an improved surface figure. Ultimately, with 107 adjusters (1.05 adjuster per square foot) a surface figure of 0.0015 inch rms was achieved. A photograph of the breadboard reflector is shown in Figure 2 with the first author, 6'0" tall, standing in front of it.

---

[1] The layup mold tool manufactured for the NASA ACTS– Advanced Communication Technology Satellite primary, 20GHz, transmit antenna was used for the fabrication of this demonstration reflector.

Figure 2. Breadboard 3.3M Diameter Reflector.

4 Conclusion

The concept of fabricating a semi–rigid and adjustable membrane reflector surface with low–cost materials and exceptional surface accuracy requirements was validated both empirically and analytically in this program. The parametric study of factors contributing to the overall surface adjustment concluded that, with a thicker membrane or more adjusters, the target accuracy for astronomical measurements at frequencies up to 900 Ghz could be met. The composites technology described in this paper is very well suited to applications in terrestrial radio astronomy with extreme tolerances for surface accuracy and dimensional stability and "down–to–earth" monetary budgets.

The adjustable membrane methodology for building large, high accuracy reflectors has been demonstrated to be technically feasible and to be a robust solution for terrestrial radio telescope applications. Because precision is applied once during the adjustment process and is not required to be maintained through the manufacturing process, the issue of applying a durable, thick, metallic coating for RF reflectivity is solved. Ultimately, the surface accuracy requirements for observations in the millimeter wavelengths can be achieved at relatively low costs compared to rigid composite panels and with greater dimensional stability than with machined metal panels.

Other possibilities for this technology are the creation of adjustable secondary reflectors for correction of wave front phase error caused by inaccurate primary reflectors, such as in a Cassegrain telescope. An adjustable reflector could be produced where the focal axis alignment can be adjusted by distorting the surface from the shape of one parabola into a new parabola with a different focal length, offset or focal axis pointing vector. A parabolic reflector could be adjusted to have many different Zernike aberration mode shapes.

5 References

1. Muser, D. Experience with Large CFRP–Radio telescope Reflectors. ESA, Spacecraft Structures and Mechanical Testing Vol. 2 p. 907–911 1991
2. Tajima, T, et. al. CFRP Surface Panels for Radio Telescope. Mitsubishi Denki Giho Vol. 56 No. 7 p. 513–17 1982

AN INSTRUMENT FOR STUDYING GALAXY EVOLUTION ON THE 9.2-METER HOBBY-EBERLY TELESCOPE

MARSHA J. WOLF,[†] GARY J. HILL,[‡] JOSEPH R. TUFTS[*]

University of Texas at Austin, [†]Astronomy Department, [‡]McDonald Observatory, 2511 Speedway, Austin, TX 78712, USA*
E-mail: [†]mwolf@astro.as.utexas.edu, [‡]hill@astro.as.utexas.edu, [*]grin@astro.as.utexas.edu

We present the description of a near infrared extension to the 9.2-m Hobby-Eberly Telescope (HET) Low Resolution Spectrograph (LRS), which will cover the wavelength range of 0.87 to 1.35 μm. The LRS-J, an upgrade to the existing LRS, replaces the optical camera with an f/1 camera optimized for the J-band. The instrument design is strongly motivated by the desire to observe galaxies at $1 < z < 2$, where the principal emission lines indicative of star formation are shifted into the J-band. The cryogenically cooled camera will be mated to the warm LRS spectrograph, which does not result in enough thermal emission background to compromise its performance since we are primarily interested in wavelengths up to 1.35 μm for redshifts up to $z = 2$. LRS-J represents a rapid and cost-effective way to enable multi-object near IR spectroscopy on an 8-meter class telescope. The camera will use a Hawaii 1024^2 HgCdTe detector array in a dewar cooled by a hybrid closed cycle helium refrigerator and LN_2 tank. A novel feature of the instrument is the use of large volume holographic (VH) grisms, which will approach efficiencies of 90% in the near IR. The complete instrument will achieve peak efficiencies in the J-band of 25% on the sky. The spectral range will be covered by 2 grisms, resulting in instrument resolving power of $R = 1400 - 1600$. Simulations of predicted instrument performance and planned galaxy evolution studies are presented.

1 Introduction

The J-band extension to the Hobby-Eberly Telescope (HET)[1] Low Resolution Spectrograph (LRS)[2,3], LRS-J[4], is primarily motivated by the desire to observe galaxies at redshifts of $1 < z < 2$. This regime is extremely important for studies of galaxy evolution, as it appears to be a transition epoch from irregular galaxy structures, which may suggest a high merger rate, to the more ordered morphologies seen locally. Unfortunately, useful spectral features of galaxies in this epoch are shifted into the J-band, beyond the sensitivity of the CCD detectors in most multi-object instruments, so this region has not been extensively studied. In this wavelength range, HgCdTe detectors have QE >50%, which makes it moderately straightforward to obtain redshifts of L^* galaxies in 1 hour integrations on the HET. Hence, important strides in our understanding of galaxy evolution can be made using our LRS-J camera in conjunction with the LRS in its multi-object mode.[a]

2   Instrument Overview

LRS-J is a simple upgrade to the existing spectrograph, as it replaces the optical camera with one optimized for the J-band. Spectral coverage for this instrument is limited to $\lambda < 1.35\mu$m for sky-limited operation due to the thermal background, because the cryogenically cooled camera will be mounted to a warm spectrograph. However, because we are primarily interested in wavelengths short of 1.35 μm for galaxies at $z = 1 - 2$, this will not compromise performance. The science focus of the instrument is well served by observations extending through the J-band. The current design accesses the entire J-band with one volume holographic (VH) grism, and the 0.85-1 μm range with a second grism. Both grisms can be carried in the LRS concurrently.

The instrument will have three observing modes covering the 0.85 to 1.35 μm wavelength region: imaging over the 4 arcmin HET field of view in z and J bands; longslit spectroscopy with slit widths ≥ 1 arcsec; and multi-object spectroscopy (MOS)[5] of up to 13 objects simultaneously with 1.3 x 15 arcsec slits on 20 arcsec centers. In spectroscopy mode the exposure times will be approximately 5 minutes and the telescope will be nodded between two positions on the slit to provide improved sky subtraction. As with the optical LRS, the imaging mode will be used for field acquisition and setup.

Optimal resolution is an important design consideration for an instrument operating in the J-band where strong atmospheric OH^- emission lines can adversely affect the signal-to-noise ratio (S/N) of observations. High spectral resolution can allow a large fraction of the resolution elements to see a lower background in between these OH^- lines, but there is a limit to how high a resolution is effective.[6] Simulations for LRS-J show that our optimum resolution is $R \sim 2000$, but that $R \sim 1600$ is adequate to reach faint limits for a majority of the pixels. A resolution of R$\sim$1600 also allows the entire J-band to be observed with one grating. Higher resolution gratings may be added to allow measurement of galaxy dynamics.

We predict an on sky peak efficiency of the whole instrument (including the HET) of 25%. Preliminary simulations indicate a limiting continuum magnitude of $J \geq 21$ in a 4 hour integration.

---

[a]The Marcario Low Resolution Spectrograph is a joint project of the Hobby-Eberly Telescope partnership and the Instituto de Astronomia de la Universidad Nacional Autonoma de Mexico. The Hobby-Eberly Telescope is operated by McDonald Observatory on behalf of The University of Texas at Austin, the Pennsylvania State University, Stanford University, Ludwig-Maximilians-Universitaet Muenchen,and Georg-August-Universitaet Goettingen.

3 Simulated Performance

Simulations are being run to estimate the performance of LRS-J. Artificial data were created with the IRAF *artdata* package. Night sky OH^- emission line intensities and continuum levels between lines were taken from Maihara[7] and Osterbrock.[8,9] An input Sb galaxy with emission lines (NGC-6181) from Kennicutt[10] was redshifted to $z = 2$, scaled to various J magnitudes, and added to the sky backgrounds. Noise was added in the following manner: 25 e^- for thermal background in a 5 minute exposure, 6 e^- detector read noise, and Poisson photon noise for each exposure. A number of 5 minute exposures were generated for two locations on the slit to simulate nodding, the sky was subtracted, and they were coadded to give 1, 2, and 4 hour total exposures. The spectra were extracted using *apall* in IRAF and the S/N was measured with *splot*.

The simulations have resulted in a limiting magnitude of $J \sim 21$ in 4 hours for 5σ per resolution element, smoothed to a resolving power of $R = 300$. Figure 1 shows an example of the data. An extracted spectrum is shown in the left panel for an Sb galaxy of $J = 21$ at a redshift of 2 for a 4 hour integration. Approximately 20% of the pixels have residual sky noise on the 5σ level due to inadequate subtraction of strong emission lines. The same spectrum is shown in the right panel after it was median filtered (corresponding to smoothing from $R = 1600$ to $R = 300$), along with the input galaxy template. We are developing masking techniques in software to block pixels containing strong OH- emission features, which should help cut down the number of spurious features in the filtered spectra, but may result in the loss of data if a galaxy emission line falls on a strong OH^- line. Thus far, atmospheric absorption has not been included in the simulation, but will be added in the next iteration.

4 Planned Science Projects

Galaxy evolution studies using LRS-J are planned for both field and cluster galaxies at redshifts of $z \sim 1$-2. A J-band galaxy survey will be conducted on fields of interest. The star formation history of field galaxies will be extended into this epoch by measuring Hα and [OII] emission lines. K-band luminosities of the galaxies will be used to normalize the star formation rates (SFRs) to galactic stellar mass, while FIR and submm observations will provide estimates on the dust mass in the galaxies. Other methods for measuring gas mass will be investigated so that the SFRs can also be normalized to gas mass. If morphological information is available, then a distribution of specific SFRs of galaxies of different types will be compiled for this epoch.

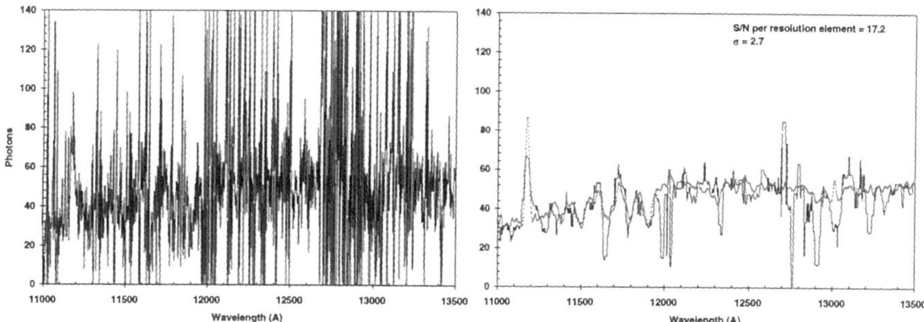

Figure 1. The left panel shows a simulated extracted spectrum of a $J = 21$, $z = 2$, Sb galaxy with 4 hours of integration time. The right panel shows the same spectrum after median filtering ($R = 1600$ to $R = 300$) with 5σ clipping. The dotted spectrum is the input galaxy template. The average S/N=17.2 for the whole spectrum, and is higher in regions with less residual sky noise.

LRS-J will also be used to verify high redshift galaxy clusters spectroscopically. New search techniques and satellites, such as Chandra, are producing high redshift cluster candidates. Clusters at redshifts of $z > 1$ are ideal targets for LRS-J.

References

1. Ramsey L.W., et al., Proc. SPIE **3352**, (1998).
2. Hill G.J., et al., Proc. SPIE **3355**, 375 (1998).
3. Hill, G.J., Proc. SPIE **4005**, 240 (2000).
4. Tufts J.R., et al., Proc. SPIE **4008**, 736 (2000).
5. Wolf M.J., et al., Proc. SPIE **4008**, 216 (2000).
6. Thompson, K.L., Mon. Not. RAS **303**, 15 (1999).
7. Maihara T., et al., PASP **105**, 940 (1993).
8. Osterbrock D.E., et al., PASP **108**, 277 (1996).
9. Osterbrock D.E., et al., PASP **109**, 614 (1997).
10. Kennicutt R.C., ApJ **388**, 310 (1992).

4. Source Counts and Counterparts of Submm Sources

STUDY OF LOCKMAN HOLE ISOPHOT SOURCES

M. YUN (NRAO/U. MASS), D. SANDERS (IFA), K. KAWARA (U. TOKYO),
Y. TANIGUCHI (TOHOKU U.), H. OKUDA (ISAS), C. CARILLI (NRAO)

To estimate the star formation history hidden from the optical, we have conducted a deep far-infrared survey at 95μm and $175\mu m$ using ISO. Multi-wavelength follow-up studies are conducted from radio to optical wavelengths in order to derive physical characteristics of these dusty starburst galaxies. These data are interpreted using a starburst spectral energy distribution model, and this model is also explored as a promising new way to infer photometric redshifts and intrinsic luminosities of the distant, optically faint submm/FIR sources.

1 Introduction

In order to estimate the star formation history hidden in the optical, we have conducted a deep far-infrared survey using the ISOPHOT instrument on the Infrared Space Observatory at 95μm and $175\mu m$[1]. The survey was made in two $44' \times 44'$ fields (LH_ROSAT, LH_NW) in the Lockman Hole where the Galactic HI column density is the lowest. About 400 and 100 far-infrared sources brighter than the 5σ detection limit have been identified at $95\mu m$ ($\geq$ 60 mJy) and $175\mu m$ ($\geq$ 100 mJy), respectively. The differential source counts clearly indicate a large excess of far-infrared sources by a factor of 10-50 over the no-evolution models based on optical data[2,3], and a large fraction of star formation may be hidden by dust even at modest redshifts ($z \leq 1$).

The analysis of the submm SCUBA source counts and inferred redshift distribution based on the radio-submm flux density ratio[4] suggests that luminous dusty galaxies may dominate the star formation history at early epochs ($z \sim 1 - 3$)[5]. As much as 80% of all SCUBA sources are too faint ($R \geq 28$, $I \geq 25$) to be studied directly even with 10m class telescopes, and the investigation of their nature may have to wait until the arrival of future facilities such as SIRTF and NGST. On the other hand, most ISOPHOT sources are expected to be at modest redshifts and readily accessible for detailed studies because the FIR dust peak redshifts beyond the $175\mu m$ band at $z \geq 1$. For example, a dusty starburst galaxy like Arp 220 with $L_{\rm IR} \sim 10^{12}L_\odot$ should have $S_{175\mu m} \sim$100 mJy and $I \sim$21 at $z \sim$0.4.

2 Multi-wavelength Observations

To explore the nature of the strong excess of far-infrared sources, we have initiated a program of multi-wavelength observations. The main objective of our project is mapping the contribution of dusty starburst systems to the overall star formation history during the dramatic decline in the star formation rate (SFR) after the apparent peak near $z \sim 1$[6]. The first step in this effort is

90

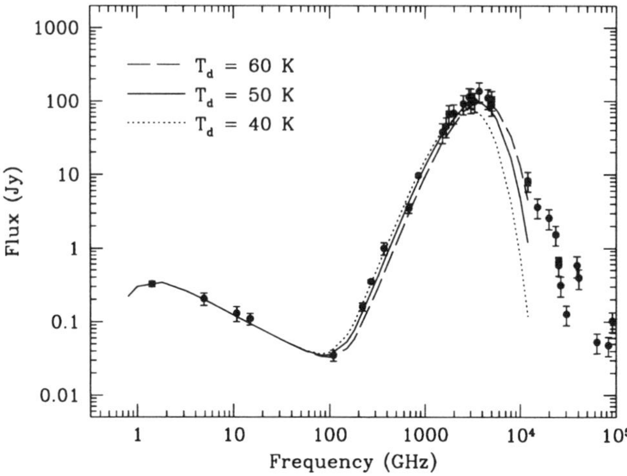

Figure 1. Radio-to-IR SED for Arp 220 is shown along with the model starburst SEDs with $T_d = 40$ K, 50 K, and 60 K are shown.

identifying the dusty starburst systems at intermediate redshifts with a deep imaging survey at $95\mu m$ and $175\mu m$ using ISOPHOT, and we have confirmed the strong evolution in the identified sources from their number counts. To quantify their evolution and the cosmic star formation history hidden by dust, we need to determine the redshifts and luminosities of the sources identified. We have begun multi-wavelength follow-up observations, first using the deep VLA radio images to secure the source identifications and later at other bands to investigate the physical properties of individual ISOPHOT sources.

3 SED Modeling and Photometric Redshifts

The spectral energy distribution (SED) for a star forming galaxy is directly proportional to the SFR between radio and far-IR wavelengths[7]:

$$S_{dust} \propto SFR(1 - e^{-\tau_{dust}})B(\nu, T_d)D^{-2}$$

$$S_{ff} \propto SFR(1 - e^{-\tau_{ff}})B(\nu, T_e)D^{-2}$$

$$S_{nth} \propto SFR\, e^{-\tau_{ff}} f_{nth}\nu^{-\alpha}D^{-2}$$

where S_{dust}, S_{ff}, and S_{nth} represent the flux density from dust, free-free, and non-thermal synchrotron emission. $B(\nu, T_d)$ and $B(\nu, T_e)$ are Planck functions for dust temperature T_d and electron temperature T_e. Submm optical depth τ_{dust} is defined so that the rising dust spectrum has emissivity of β but becomes optically thick and resembles a blackbody spectrum near 150 μm. Free-free optical depth τ_{ff} provides the floor for the spectral minimum near 100 GHz and a break in the synchrotron power-law near ≤ 1 GHz. The

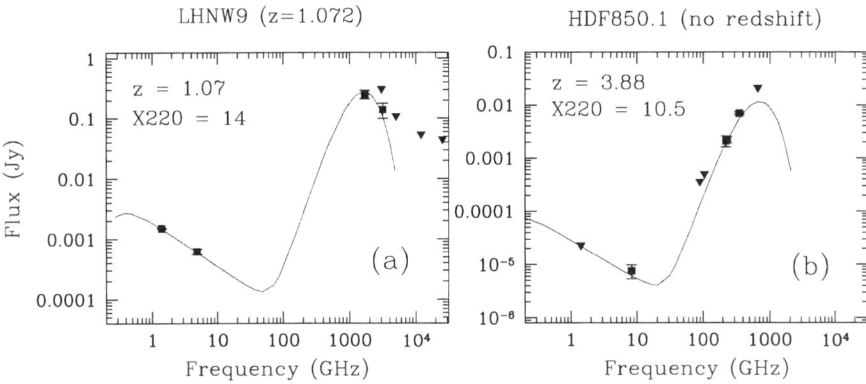

Figure 2. (a) Starburst SED modeling of the $z = 1.07$ ISOPHOT sources LHNW9. (b) Photometric redshift analysis for the brightest SCUBA Sources in Hubble Deep Field (HDF).

intrinsic synchrotron power-law index is assumed to be $\alpha \sim 0.7$.

The radio-to-IR SED data and the best fit models for the prototypical starburst galaxy Arp 220, shown in Figure 1, nicely demonstrate that this starburst SED model provides an excellent description of the observed SED despite its simplicity. For Arp 220, the T_d=50K and β=1.5 model offers a good overall fit. One way of utilizing this model is deriving IR luminosity and starburst rate for starburst galaxies with known redshift. For example, the observed SED and the best fit starburst model for the $z = 1.072$ ISOPHOT source LHNW9 are shown in Figure 2a. Even though no submm measurements are available, the combination of the rest frame radio and FIR measurements can be used to infer that this galaxy has an intrinsic FIR luminosity that is about 14 times larger than Arp 220.

As a galaxy is placed further and further away, its entire SED shifts to the bottom (fainter, because of D^{-2}) and to the left (lower frequency, Doppler shift). This generic behavior for a starburst galaxy and the resulting change in the apparent spectral index between 1.4 GHz and 850 μm has been pointed out as a redshift indicator by Carilli & Yun[4]. By analyzing the observed SEDs of the 17 starburst galaxies in the local universe, we have determined that T_d = 50 K and dust emissivity $\beta = 1.5$ offers an excellent template for the photometric redshift determination[8]. Using this template, we can estimate the redshifts more accurately by fitting the entire SED. As in most photometric redshift techniques, redshift information comes from distinct spectral features such as the sharp rise in the dust spectrum and the dust peak near the rest frequency of 100 μm. The radio synchrotron measurements help set the vertical scale with respect to the dust spectrum, i.e. SFR.

Redshifting a dust spectrum is exactly equivalent to lowering dust temperature, and a departure in T_d from the template SED translates directly to a redshift error. While inherently subject to a variation, the magnitude of uncertainty due to the spread in T_d is well understood quantitatively: $\Delta z = \frac{\Delta T_d}{T_d}$. If the characteristic T_d is higher or lower by 10 K (e.g. Figure 1), the resulting error in the photometric redshift is only 0.2!

When we apply this photometric redshift technique to the two well studied SCUBA galaxies SMM 02399$-$0134 ($z = 2.80$) SMM 14011+0252 ($z = 2.57$), we derive photometric redshifts of 2.46 & 2.62, respectively. The differences from the spectroscopic redshifts are somewhat larger than expected from the uncertainty in dust temperature alone, but these are extremely encouraging results. The biggest uncertainty arises from poor sampling of the SEDs. Even a good set of reliable measurements may offer a limited utility in resolving the z-SFR ambiguity if they are clustered along a single SED feature such as the rising part of the dust SED (e.g. SCUBA bands only) or along the radio continuum feature. On the other hand, because we are fitting the SED features changing over logarithmic scales, only a few well placed SED data points are needed to derive the redshifts[4].

A great promise of this photometric redshift technique is in estimating the redshifts for optically faint submm galaxies, which are the majority of the SCUBA sources. In Figure 2b, the best solution satisfying all upper limits and producing the minimum χ^2 for our starburst SED template is shown for the brightest SCUBA source in the HDF (HDF850.1)[9]. Our photometric redshift analysis suggests that this galaxy has a high likelihood of having a redshift near $z \sim 3.9$ with an intrinsic IR luminosity of about 10 times that of Arp 220. The uncertainty in the inferred redshift is too large ($\Delta z \geq 0.3$) for a CO search using existing instruments, but a spectroscopic verification of HDF850.1 and other optically faint submm galaxies should become possible using future broadband instruments on the GBT and the LMT.

References

1. Kawara, K., Sato, Y., Matsuhara, H., et al. 1998, A&A, 336, L9
2. Guiderdoni, B. et al. 1998, MNRAS, 295, 877
3. Matsuhara et al. 2000, A&A, in press (astro-ph/0006444)
4. Carilli, C., & Yun, M.S. 1999, ApJ, 513, L13
5. Barger, A., Cowie, L.L., & Richards, E.A. 2000, AJ, 119, 2092
6. Madau, P., et al., 1996, MNRAS, 283, 1388
7. Condon, J. 1992, ARAA, 30, 575
8. Yun, M.S., & Carilli, C. 2001, in preparation.
9. Hughes, D. H., et al. 1998, Nature, 394, 241

THE BRIGHTER SIDE OF SUB-MM SOURCE COUNTS: A SCUBA SCAN-MAP OF THE HUBBLE DEEP FIELD

COLIN BORYS[1], SCOTT CHAPMAN[2], MARK HALPERN[1]
DOUGLAS SCOTT[1]

[1] *Department of Physics & Astronomy, University of British Columbia, Vancouver BC CANADA V6T 1Z1*
E-mail: borys@physics.ubc.ca

[2] *Observatories of the Carnegie Institution of Washington, Pasadena CA 91101 USA*

We present an 11 × 11 sq. arcmin map centred on the Hubble Deep Field taken at $850\,\mu$m with the SCUBA camera on the JCMT. The map has an average one-sigma sensitivity to point sources of about 2.3 mJy and thus probes the brighter end of the sub-mm source counts. We find 7 sources with a flux greater than $9\,$mJy ($\sim 4\sigma$), and therefore estimate $N(> 9\,\mathrm{mJy}) = 208^{+90}_{-72}$ degree^{-1}. This result is consistent with work from other groups, but improves the statistics at the bright end, and is suggestive of a steepening of the counts.

1 Introduction

Observations using the Submillimetre Common User Bolometer Array (SCUBA[1]) have been revolutionizing our understanding of the importance of dust in galaxies at high redshift. Already several very deep integrations have been carried out on single SCUBA fields down to the confusion limit. In order to learn more about source counts (and hence to constrain models), the next step is to search for brighter objects over somewhat larger fields.

The population of bright sub-mm sources is currently not well understood. Current models[2,3,4,5] for source counts in the sub-mm have been able to account for the observed sources by invoking evolution which follows the $(1+z)^3$ form required to account for IRAS galaxies at 60 μm, and the powerful radio-galaxies and quasars[6]. Euclidean models with no evolution have a slope of roughly -1.5 ($N_s \propto S_\nu^{-3/2}$), which cannot possibly account for the sources observed. With reasonable evolution (in IRAS-motivated models), the counts steepen sharply at the 10's of mJy sensitivity level to roughly $S_\nu^{-2.6}$. At the source detection level of previous work, typically 5 mJy, there is little variation between the models, and we are well into the steep counts regime.

However, at the 10–30 mJy sensitivity level various evolutionary models[7] show more parameter dependence. Given a possibly wide range of galaxy types contributing to the source counts at the bright end, the actual counts may deviate from a simple parametric model dramatically. Furthermore, cosmological, as well as evolutionary parameters, play a role here.

2 Data collection and analysis

The standard scan-map observing strategy is to use multiple chop throws in two fixed directions on the sky so that an Emerson deconvolution technique can be applied. This may be appropriate for galactic plane mapping, where structure appears on all scales, but we seek a simpler approach which is optimized for finding point sources. The disadvantage of the standard approach is that the off-beam pattern gets diluted (making it more challenging to isolate faint sources), and the map noise properties are more difficult to understand. Our approach was to use a single chop direction and throw fixed on the sky. The result is a map that has, for each source, a positive and negative signal.

A total of 61 scans were obtained, 31 in early 1998 and 30 more a year later. The data were first analysed using SURF, but because the data were taken in a non-standard mode, we found it necessary to write custom software. In this way, we were able to isolate and remove large scale features in the maps, and estimate the per pixel noise level by a careful accounting of the per-bolo noise and the frequency at which it sampled a particular pixel.

Sources were found by using a model for the dual beam pattern and fitting it, in a least-squares sense, to each pixel in the map. A total of 7 sources were found that had a peak to error-of-fit ratio greater than 4. We also rotated the dual beam pattern by ninety degrees and used that as a model; Only 2 sources were found, one of which is associated with the brightest source in the map. Monte-Carlo simulations suggest that the other false positive is not unexpected. As an additional check, the 1998 data were compared with the 1999 data to ensure that each source was evident in both halves of the data.

The calibration was determined by fitting the beam model to observations of standard calibrators taken during the run. We checked our procedure by re-analysing the Barger[8] et al. HDF/radio fields, which are a set of jiggle maps within the HDF flanking fields. In fact this led to a correction in our original flux estimates. Fig.1 presents our final map, and an updated source count plot is given in Fig.2.

3 Conclusions and followup work

We have significantly improved statistics at the $\sim 10\,\mathrm{mJy}$ level, and suggest there is some indication of a steepening of the counts there. Because these sources are brighter than typical SCUBA detections, they will be relatively

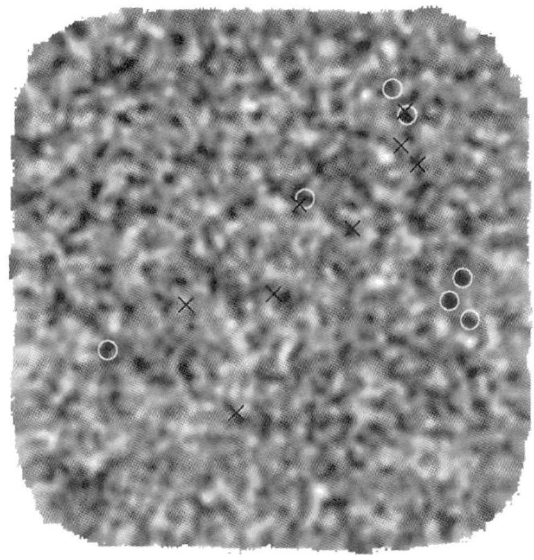

Figure 1. The 850 μm HDF map. The map size is roughly 12×12 sq. arcmin, although only the central 11×11 sq. arcmin was used in order to avoid the noisy edges. The chop is 40 arcsec and is roughly east-west. The circles outline the 7 sources detected in our survey and the dark crosses highlight sources found in the Barger et al. (2000) and Hughes et al. (1998) work. Since these surveys went deeper, not all can be detected in the scan map. They also cover a smaller area, and hence cannot see several of our sources.

easy to follow up at other wavelengths. Preliminary analysis already indicates good correlation with μJy radio sources, but little indication of optical counterparts (as found in other studies).

Although not discussed here, we are also investigating sources in the 450 μm scanmap. However, this is more arduous given the lower sensitivity of SCUBA with this filter. Finally, a similar analysis is being conducted on data taken of 50 sq. arcmin of the Groth strip/Westphal region.

Acknowledgments

We would like to thank Amy Barger for access to her data before it was available on the JCMT public archive, and Doug Johnstone for several useful conversations. We are also grateful to the staff at the JCMT, particularly Tim Jenness and Wayne Holland who provided invaluable advice.

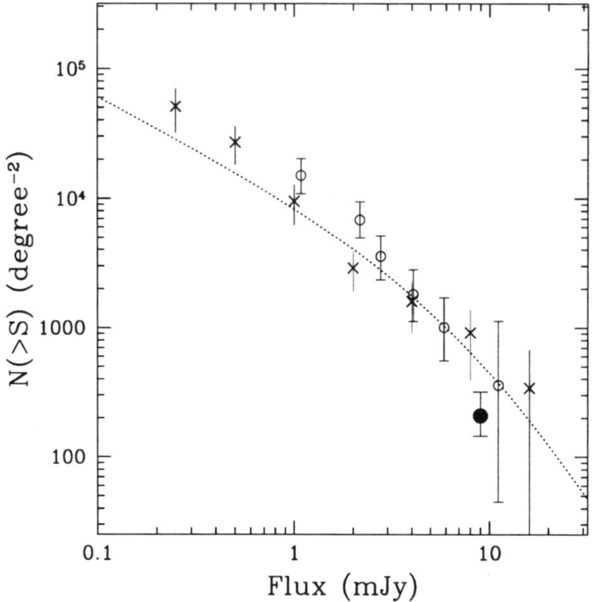

Figure 2. The $850\,\mu$m source counts. The crosses are data from Blain et al. (1999) and the open circles that from a recent cluster survey by Chapman et al. (2000) The dashed line is a simple two power-law fit to the previous counts. Our new estimate, with Bayesian 68% confidence limits, is given by the solid circle.

References

1. Holland, W.S. *et al*, *MNRAS* **303**, 659 (1999).
2. Blain, A.W., *MNRAS* **295**, 92 (1998).
3. Eales S., Edmunds M.G., *MNRAS* **280**, 1167 (1996).
4. Blain A.W., Longair M.S., *MNRAS* **279**, 847 (1996).
5. Fall S.M., Charlot S., Pei Y.C., *ApJ* **464**, L43 (1996).
6. Dunlop J.S., Peacock J.A., *MNRAS* **247**, 19 (1990).
7. Guiderdoni B. *et al*, *MNRAS* **295**, 877 (1998).
8. Barger, A.J., Cowie, L.L., Richards, E.A., *AJ* **119**, 2092 (2000).
9. Hughes D.H. *et al*, *Nature* **394**, 241 (1998).
10. Blain A.W. *et al*, ASP Conf. Ser. **193**, 246 (1999).
11. Chapman, S.C. *et al*, *MNRAS*, submitted, astro-ph/0009067 (2000).

SUB-MM COUNTERPARTS TO LYMAN-BREAK GALAXIES

SCOTT CHAPMAN

*Observatories of the Carnegie Institution of Washington, Pasadena, CA
91101, U.S.A.*
E-mail: schapman@ociw.edu

DOUGLAS SCOTT, COLIN BORYS, MARK HALPERN

*Department of Physics & Astronomy, University of British Columbia, Vancouver,
B.C. V6T 1Z1, Canada*

We summarize the main results from our SCUBA survey of Lyman-break galaxies
(LBGs) at $z \sim 3$. Analysis of a new large sample of LBGs reveals *some* 850
micron emission is likely present in our high star formation rate (SFR) LBG sample
(0.7 mJy). Known populations of LBGs could therefore be contributing a large
portion of the far-IR Background. The detection of the LBG, Westphal-MM8, at
1.9 mJy suggests that deeper observations of individual LBGs in our sample could
uncover detections at similar levels, consistent with our UV-based predictions. By
the same token, many sub-mm selected sources with $S_{850} < 2$ mJy could be LBGs.
The data is also consistent with the far-IR/β relation holding at $z = 3$.

1 Introduction – Selecting sub-mm luminous LBGs

The Sub-millimeter Common User Bolometer Array (SCUBA[1]) on the 15-m
JCMT[a] has sparked a revolution in our understanding of dust obscured star
formation and active galactic nuclei. However, going back several years to the
initial discovery of sub-mm sources in the field[2], we were rather naive about
the nature of sub-mm selected sources. It seemed at the time perfectly reasonable that the high star formation rate tail of the known $z \sim 3$ LBG population
should allow us to quickly pre-select and observe some large fraction of the
bright SCUBA sources.

The R-band for $z \sim 3$ galaxies (rest frame UV continuum) provides a
direct measure of the light from young, hot stars. The spectral slope through
the rest frame UV continuum allows a determination of the extinction of this
light due to interstellar dust, and thereby prescribes a dust correction to the
star formation rate (SFR). Thus the goal of our project was to choose those
few LBGs exhibiting a combination of a bright R-magnitude and a steep
$g - R$ slope (large dust correction) resulting in a sample of galaxies thought

---

[a]The JCMT is operated by the Joint Astronomy Centre on behalf of the United Kingdom
Particle Physics and Astronomy Research Council (PPARC), the Netherlands Organisation
for Scientific Research, and the National Research Council of Canada.

to have high SFRs ($> 200\,M_\odot/\mathrm{yr}$) and thus be detectable in the sub-mm with SCUBA. It is important to note that these objects were not just a few outliers with extremely large dust correction factors, but were also amongst the most intrinsically luminous of the LBG population.

1.1 An initial sample of high SFR LBGs

The results of our initial study have been presented in Chapman et al. (2000)[3], with further implications explored in Adelberger & Steidel (2000)[4]. From an original sample of 16 LBGs chosen for followup with SCUBA, only 8 turned out to have the sought properties (high expected SFR) upon more detailed optical analysis. Predicting the 850 micron flux density (S_{850}) for LBGs from the UV continuum was accomplished by first employing the empirical farIR/β relation for local starburst galaxies[5] to estimate the total farIR luminosity from the UV slope (β). We then identified the S_{850} point through an empirical measure of the typical spectral energy distribution (SED) for star forming galaxies in the farIR/sub-mm region, extracted from a large sample of local LIRGs and ULIRGs[4,3]. A clear uncertainty in this prescription is the paucity of high-z star forming objects with sub-mm measurements available to validate our prediction recipe.

SCUBA observations during observing runs in winter 1998 yielded only one clear detection in our sample, Westphal-MMD11. This object in particular was found to be rather extraordinary in many of it's properties, and will be discussed in the following section. The UV-based predictions of sub-mm flux density were typically too high for the sample on the whole, except for Westphal-MMD11 which has considerably more sub-mm flux than expected. However, large errors bars and a relatively small sample of objects made it difficult to extract robust conclusions from the study. The results of this study are summarized in Figure 1. However, new evidence as presented below suggests that this original study may have been a somewhat misleading result.

2 Sub-mm detected LBGs at $z \sim 3$

2.1 Westphal-MMD11

Westphal-MMD11 currently represents the highest redshift source ($z = 2.98$) detected with SCUBA, which is not an AGN. Upon subsequent optical study, Westphal-MMD11 seems more akin to HR10[6], the dusty starbursting ERO, than what one might consider a local analog of LBGs such as M82. It is luminous in the near-IR with $R - K = 4.5$, almost 2 magnitudes larger in $R - K$ than the median for the LBG population[7]. With $S_{850} = 5.5\,\mathrm{mJy}$ and

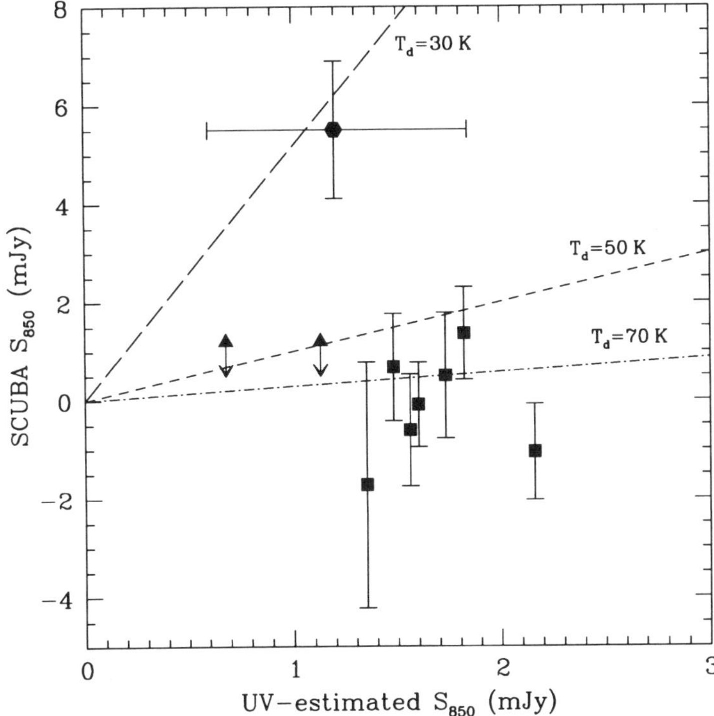

Figure 1. The predicted vs. measured sub-mm flux density for our initial sample of eight LBGs. The dust temperature dependence of our predictions is shown for 30, 50 and 70 K. Also plotted are 2 sources from the HDF (triangles).

L_{FIR} close to $10^{13} h_{50}^{-2} L_\odot$, the question perhaps is why is it visible in the UV at all? A recent near-IR spectrum taken with NIRSPEC on the Keck telescope[8] reveals a double peaked line profile, with continuum emission only present under one peak. This suggests geometrical effects involving large amounts of dust are likely at work, perhaps similar to the merging ULIRG, Arp220.

2.2 SMMJ14011

The lensed galaxy SMM J14011+0252[9,10] is the only sub-mm selected object clearly identified as a high redshift ($z = 2.565$) galaxy without an obviously active nuclues[11]. Followup optical photometry in a filter set matched to the

Steidel et al. (1999)[12] LBG survey[4] revealed that the restframe-UV SED of SMMJ14011 is actually indistinguishable from Westphal-MMD11, yet their optical SEDs are quite different. Within the SMM J14011+0252 system, interferometry observations[10] reveal that the dominant sub-mm emission resides in a red component (J1), with the dominant UV emission appearing in a neighboring component (J2). At face value, this questions the validity of the unobscured star formation rate prediction based on the UV. The UV properties of a spatially separated component should have little bearing on the dust and star formation in J1. On the other hand the empirical correlation between β and $L_{bol,dust}/L_{UV}$ is rather mysterious – it is difficult to understand why this correlation should exist at all, but local observations suggest that it does, and the scatter in the relation may relate a variety of processes involved in starbursting galaxies at various evolutionary stages.

2.3 Westphal-MM8

Identified as a LBG in the surveys of Steidel et al. (1999)[12], Westphal-MM8 possesses a rest-frame UV spectrum which indicates an exceptionally large SFR ($> 100 M_\odot/yr$). However, the infrared colors and overall properties do not single this object out as unusual for the population in any other respects. Although still a relatively extreme object, Westphal-MM8 represents one of the faintest galaxies ever detected with SCUBA ($S_{850} = 1.9 \pm 0.5\,\mathrm{mJy}$), comparable to the faintest objects in the Hubble Deep Field SCUBA observation of Hughes et al. (1998)[15]. Simply scaling the observed submillimetre photometry from that of Arp220 implies that MM8 has an infrared luminosity of $L_{IR} \sim 4 \times 10^{12}\ h_{50}^{-2} L_\odot$. Although original results[3,4] were somewhat pessimistic about the sub-mm detection of LBGs, this measurement of MM8 gives us renewed confidence that we may be able to pre-select those extreme LBGs emitting in the S_{850}=1-2 mJy range.

3 A larger sample of high-SFR LBGs

The case of Westphal-MM8 reveals the inherent difficulty in detecting even the most sub-mm luminous members of the LBG population with SCUBA. Clearly spending two JCMT shifts per object is not an option for studying the properties of a large sample of LBGs.

A new and larger sample consisting of 33 LBGs has now been observed with SCUBA. The sample breaks down as follows: 8 high SFR LBGs from our original sample, 12 red LBGs ($R - K > 3.5$) which could be 'like' W-MMD11, and 13 new LBGs similar to the original sample of large predicted

SFR. Some of these objects are selected from fields with better photometry than the original surveys, and photometric errors could conceivably be less of an uncertainty. Still, only 2 clear detections emerge (W-MMD11 and W-MM8), although certain sub-samples exhibit clear detections in their average flux density.

We summarize the average properties of this new sample as follows:

$S_{850} = 0.7 \pm 0.3$ mJy (0.6 ± 0.2 weighted by the inverse variance)

$S_{450} = 10.4 \pm 5.8$ mJy (2.9 ± 2.2 weighted by the inverse variance)

These numbers include MMD11, as the numbers are small enough that we do not yet know if MMD11 is highly unusual, or merely the tail end of the red LBG population. Several other red LBGs are marginally detected with $S_{850} \sim$ 3 mJy. The S_{850} predicted from the UV for this sample is approximately 1.5 mJy and is thus is not drastically in conflict with the measured flux. This result indicates that the far-IR/β relation appears at least consistent at $z = 3$ with what we observe locally.

3.1 Far-IR background

For an $\Omega = 0.3$, $\Lambda = 0.7$ universe, the average dust luminosity of our SCUBA observed LBG sample is $L_{bol}(\text{dust}) = 2.7 \pm 1.2 \times 10^{11} L_{\odot}$. The average $R(\text{AB})=23.9$ implies $L_{UV} = 4.6 \times 10^{10} L_{\odot}$. The average *obscuration* for the sample, $< L_{dust}/L_{UV} >= 5.9 \pm 2.6$ (or 4.8 ± 1.6 in the inverse variance weighted mean). UV-based predictions[4] require $< L_{dust}/L_{UV} >= 6$ to recover the bulk of the 850 micron background.

Although submm sources > 6 mJy appear largely distinct from the LBG population[13,14], they form a relatively small amount of far-IR background. However, deep sub-mm observations in the HDF[15,16] suggest that $S_{850} \sim$ 2 mJy sources are observed which have no obvious optical counterparts and would not form part of typical ground-based optical surveys. Thus it remains to be seen just what properties the bulk of these crucial $S_{850} \sim 1 - 2$ mJy sources actually have.

4 Conclusions

We find that SCUBA observations of a large sample of LBGs are consistent with *some* 850 micron emission in $z = 3$ LBGs (0.7 mJy). This is consistent with known populations of LBGs contributing a large portion of the far-IR background. A deep observation detecting the LBG W-MM8 at 1.9mJy suggests that individual LBGs in our sample could in fact possess such an S_{850} consistent with our UV-based predictions. By the same token, many sub-mm selected sources with $S_{850} < 2\,\text{mJy}$ could be LBGs. The data is also consistent with the far-IR/β relation holding at z=3.

Acknowledgments

We thank our collaborators, C. Steidel, K. Adelberger (Caltech) S. Morris (DAO), M. Pettini (Cambridge) M. Dickinson, M. Giavalisco (STSCI) for helping to put all the pieces together in this complicated puzzle. SCC acknowledges travel support from Carnegie Observatories to attend this conference.

References

1. Holland, W.S. *et al*, *MNRAS* **303**, 659 (1999).
2. Smail, I., Ivison, R.J., Blain, A.W., *ApJ* **490**, L5 (1997).
3. Chapman S.C., Scott D., Steidel C. *et al*, *MNRAS*, in press (2000).
4. Adelberger, K.L., Steidel, C.C., *ApJ*, in press (2000).
5. Meurer G.R., Heckman T.M., Calzetti D., *ApJ* **521**, 64 (1999).
6. Dey, A., *et al*, *ApJ* **519**, 610 (1999).
7. Shapley A., *et al*, *ApJ*, in preparation (2000).
8. Steidel, C.C., *et al*, *ApJ*, in preparation (2000).
9. Smail, I., Ivison, R.J., Blain, A.W., Kneib, J.-P., *ApJ* **507**, L21 (1998).
10. Frayer, D.T., *et al*, *ApJ* **514**, L13 (1999).
11. Ivison, R.J., *et al*, *MNRAS* **315**, 209 (2000).
12. Steidel, C.C., *et al ApJ* **519**, 1 (2000).
13. Barger A.J., Cowie L, Richards E., *AJ* **119**, 2092 (2000).
14. Chapman S.C., Richards E., Lewis G., *Nature*, submitted (2000).
15. Hughes, D.H., *et al*, *Nature* **394**, 241 (1998).
16. Peacock J., *et al*, *MNRAS*, in press (2000).

CAN DUSTY LYMAN BREAK GALAXIES PRODUCE THE SUBMILLIMETER COUNTS AND BACKGROUND? LESSONS FROM LENSED LYMAN BREAK GALAXIES

PAUL P. VAN DER WERF, KIRSTEN KRAIBERG KNUDSEN, IVO LABBÉ
AND MARIJN FRANX

Leiden Observatory, P.O. Box 9513, NL - 2300 RA Leiden, The Netherlands
E-mail: pvdwerf,kraiberg,ivo,franx@strw.leidenuniv.nl

Can the submillimeter counts and background be produced by applying a locally derived extinction correction to the population of Lyman break galaxies? We investigate the submillimeter emission of two strongly lensed Lyman break galaxies (MS 1512+36-cB58 and MS 1358+62-G1) and find that the procedure that is used to predict the submillimeter emission of the Lyman break galaxy population over-predicts the observed 850 μm fluxes by up to a factor of 14. This result calls for caution in applying local correlations to distant galaxies. It also shows that large extinction corrections on Lyman break galaxies should be viewed with skepticism. It is concluded that the Lyman break galaxies may contribute to the submillimeter background at the 25 to 50% level. The brighter submillimeter galaxies making up the rest of the background are either not detected in optical surveys, or if they are detected, their submillimeter emission cannot be reliably estimated from their rest-frame ultraviolet properties.

1 Submillimeter emission from Lyman break galaxies

Measurements of the cosmic star formation rate density (SFRD) based on surveys for Lyman break galaxies (LBGs) are affected by extinction, and attempts to correct for this effect lead to a substantial upwards revision of the SFRD. [1] Since the light absorbed in the ultraviolet (UV) is reradiated in the far-infrared (FIR), LBGs affected by extinction must emit FIR radiation, which will contribute to the submillimeter background.

The expected submillimeter emission from the population of LBGs has recently been estimated based on the observed (if not entirely understood) correlation between the spectral index β in the UV (defined by the relation $f_\lambda \propto \lambda^\beta$) and the ratio of FIR to UV flux in a sample of local galaxies. [1] Applying this relation to the LBG population, Adelberger & Steidel [2] found that the submillimeter counts and integrated background can be accounted for. As noted by these authors, these estimates are uncertain, since the validity of the β-FIR/UV correlation at high redshift has not been established, and the distribution of β values and its dependence on magnitude, and the luminosity function at faint magnitudes are poorly constrained. It is therefore necessary to verify this analysis by direct submillimeter observations of LBGs.

Table 1. Predicted (from UV color and magnitude) and observed submillimeter emission of lensed LBGs. Luminosities (not corrected for gravitational amplification) have been derived for $H_0 = 50$ km s^{-1} Mpc^{-1} and $q_0 = 0.5$. The other results do not depend on cosmology. Upper limits represent 3σ.

	MS 1512+36-cB58	MS 1358+62-G1
amplification factor	~ 50	~ 10
β (observed)	-0.74 ± 0.1	-1.63 ± 0.1
A_{1600} (predicted)	3.0 ± 0.2	1.2 ± 0.2
$S_{\mathrm{FIR}}/S_{1600}$ (predicted)	18	2.4
L_{FIR} (predicted)	$3.9 \cdot 10^{13}$ L$_\odot$	$3.3 \cdot 10^{12}$ L$_\odot$
S_{850} (predicted)	58 mJy	5 mJy
S_{850} (observed)	4.2 ± 0.9 mJy	< 4 mJy
L_{FIR} (derived)	$2.8 \cdot 10^{12}$ L$_\odot$	$< 2.6 \cdot 10^{12}$ L$_\odot$
$S_{\mathrm{FIR}}/S_{1600}$ (derived)	1.3	< 1.9
A_{1600} (derived)	0.8	< 1.0

2 Lessons from strongly lensed Lyman break galaxies

The predicted 850 μm fluxes for most LBGs based on their UV properties[2] are 1 mJy or less, which is too faint for current instrumentation. Submillimeter observations of individual LBGs are significantly easier if strongly lensed LBGs are targeted. We have used SCUBA on the JCMT to observe two strongly lensed LBGs: the object cB58 at $z = 2.72$ lensed by the cluster MS 1512+36[3] and the object G1 at $z = 4.92$ lensed by the cluster MS 1358+62.[4] Rest-frame UV colors indicate significant reddening in both of these objects.[5,6] For both galaxies, the method of predicting the FIR emission based on the UV property β[7,6] implies strong 850 μm emission (Table 1). The results of our SCUBA measurements are given in Table 1. The object cB58 is detected at the 4.7σ level; the object G1 was not detected. In both cases the procedure of predicting the submillimeter flux from the observed color and magnitude in the rest-frame UV[2] overpredicts the submillimeter emission. The magnitude of the discrepancies is illustrated in Fig. 1. For G1 the discrepancy is not significant, given the scatter in the β-FIR/UV relation, but the factor 14 discrepancy for cB58 is highly significant.

Since this discrepancy is so large, it cannot be attributed to observational uncertainties. However, an element of uncertainty in the analysis is introduced by differential lensing, if the effective amplification factor of cB58 in the UV is a factor of 10 higher than that in the FIR. A large discrepancy can be introduced only if most of the UV emission comes from the most strongly

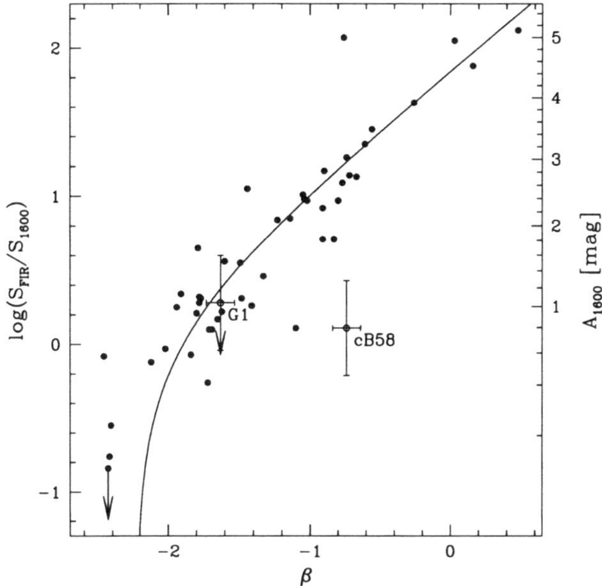

Figure 1. The relation between $S_{\mathrm{FIR}}/S_{1600}$ and β for local UV-selected galaxies (filled circles) with the best-fitting parametrization, [1] and the positions of cB58 and G1 with respect to this relation.

amplified portions of the source near the caustic, while most of the FIR emission comes from more weakly lensed regions. This situation would require a very different distribution of FIR and UV emission. However, an extinction correction based on UV properties of one region of a galaxy will not be able to predict the submillimeter emission in a completely different region of the system. Therefore, if differential lensing plays a major role, the physical basis of using the β-FIR/UV correlation disappears. For cB58, a lensing model based on new HST data shows that the amplification factor is at least a factor of 5 at every position, and that the intensity-weighted amplification factor in the UV is about a factor of 25 (on both sides of the fold arc, so that the total amplification is approximately a factor 50). Thus differential lensing can account for a discrepancy of at most a factor of 5, but probably much less, since UV and FIR emission should have a similar morphology for the β-FIR/UV correlation to work.

3 Discussion and conclusions

These results demonstrate that attempts to produce the submillimeter counts
and background based on an extinction correction applied to the LBG population [2] are fraught with considerable uncertainty, and fail in the case of the
two lensed LBGs discussed here. While our sample is small, the results argue
against the validity of the low redshift β-FIR/UV correlation in the case of
LBGs. Even if the local correlation were valid for high redshift galaxies as
well, it still would not be able to produce the brighter submillimeter galaxies.
These objects have luminosities putting them in the class of the ultraluminous
infrared galaxies (ULIGs), which do not follow the β-FIR/UV correlation, as
shown by recent HST-STIS data of a sample of nearby ULIGs. Since these
galaxies already account for $\sim 50\%$ of the submillimeter background, it is not
possible for the LBGs to produce a dominant fraction (let alone all) of the
submillimeter background. A more likely situation is that the LBGs account
for 25 to 50% of the submillimeter background as indicated by the faint structure in the HDF at 850 μm, [8] but that the dominant part of the submillimeter
background is made by a small number of ULIGs, which are either not detected in LBG surveys, or if they are detected, cannot be reliably corrected for
extinction, in the same way that local ULIGs cannot be extinction corrected
based on UV data.

In summary therefore, these results support the view that the brighter
850 μm galaxies making at least 50% of the submillimeter background form
a population that cannot be reliably reproduced by extinction corrections
applied to the LBG population.

References

1. G.R. Meurer, T.M. Heckman and D. Calzetti, ApJ **521**, 64 (1999)
2. K.L. Adelberger and C.C. Steidel, astro-ph/0001126
3. H.K.C. Yee, E. Ellingson, J. Bechtold, R.G. Carlberg and J.-C. Cuillan-
   dre, AJ **111**, 1783 (1996)
4. M. Franx, G.D. Illingworth, D.D. Kelson, P.G. van Dokkum and K.-V.
   Tran, ApJ **486**, L75 (1997)
5. E. Ellingson, H.K.C. Yee, J. Bechtold and R. Elston, ApJ **466**, L71 (1996)
6. B.T. Soifer, G. Neugebauer, M. Franx, K. Matthews and G.D. Illingworth, ApJ **501**, L171 (1998)
7. M. Pettini, C.C. Steidel, K.L. Adelberger, M. Dickinson and M. Giavalisco, ApJ **528**, 96 (2000)
8. J.A. Peacock *et al*, astro-ph/9912231

5. Molecular Emission Lines at High Redshift

HIGH REDSHIFT CO LINE EMISSION: PERSPECTIVES

F. COMBES

DEMIRM, Observatoire de Paris,
61 Av. de l'Observatoire, F-75 014, Paris, France
E-mail: francoise.combes@obspm.fr

Although about a dozen high redshift ($z > 2$) starburst galaxies have been recently detected in the CO lines, spectroscopic detections of molecular gas of very young galaxies are still very difficult. The CO lines are usually optically thick, which limits greatly the increase of emission with redshift, as observed for the dust continuum. This paper discusses the significant progress that future instruments (LMT, ALMA) will make in this area. The computations are based on an extrapolation of our understanding of starburst galaxies at lower redshift.

1 Introduction

One of the breakthroughs in observational cosmology, due to recent progress on faint galaxies, has been the inventory of the amount of star formation at every epoch[24]. The co-moving star formation rate appears to increase like $(1+z)^4$ from $z = 0$ to $z = 1$, and then decreases again to the present value at $z = 5$. But this relies on the optical studies, i.e. on the UV-determined star forming rates in the rest-frame. If early starbursts are dusty, this decrease could be changed into a plateau[18,5]. To tackle this problem, independent information must be combined in a coherent picture, such as that coming from the far-infrared and millimeter domains: the cosmic IR and submm background radiation discovered by COBE [30,21] yields an insight into the previous global star-formation history of the Universe, and the sources discovered at high redshift in the millimeter continuum and lines yield information on the structure of the past starbursts[38,22,2,19].

This review focuses on the molecular content of galaxies that we can deduce from CO lines. Although the submillimeter continuum is more easy to detect at high redshift, it cannot without ambiguity trace the evolution of star formation as a function of redshift, because of identification problems, unknown redshifts, and the unknown contribution of AGN activity. After summarising the present state of knowledge concerning CO emission lines, the future surveys that will be conducted with the next generation of millimeter instruments will be described.

2   CO Detections at High Redshift

The detection of highly redshifted millimeter CO lines in the hyperluminous $z = 2.28$ object IRAS 10214+4724 has opened this new field to explore the history of star formation, and its efficiency[7,36]. Although the first enormous derived H_2 mass has now been revised (the source is considerably amplified by lensing), it is still surprising to find such huge amounts of CO molecules, especially since the gas is expected to have lower metallicity at high z. But theoretical calculations have shown that in a violent starburst, the metallicity could reach solar values very quickly[14].

Today, more than a dozen objects have been detected in CO lines at high redshift[3,4,25,26,16,23]. They are often gravitationally amplified, either being multiply imaged by a strong lens, like the Cloverleaf quasar H1413+117 at $z = 2.558$ the lensed radiogalaxy MG0414+0534 at $z = 2.639$ or the possibly magnified quasar BR1202-0725 at $z = 4.69$; or they are more weakly amplified by a foreground galaxy cluster, like the submillimeter-selected hyperluminous galaxies SMM02399-0136 at $z = 2.808$, and SMM 02399-0134 at $z = 1.062$. Often several high-J CO lines are detected, revealing the high temperature of the gas (typical of a starburst at $\sim 60K$). Higher temperatures are rare, as in the magnified BAL quasar APM08279+5255, at $z = 3.911$, where the gas temperature derived from the CO lines is $\sim 200K$, maybe excited by the quasar[12]. Scoville et al. (1997b) reported the CO detection of the first non-lensed object at $z = 2.394$, the weak radio galaxy 53W002, and Guilloteau et al. (1997) detected the radio-quiet quasar BRI 1335-0417, at $z = 4.407$, which has no direct indication of lensing. If the non-amplification is confirmed, these objects would contain the largest molecular contents known ($8 - 10 \cdot 10^{10}$ $M_\odot$ with a standard CO/H_2 conversion ratio, and even more if the metallicity is low). The derived molecular masses are so high that H_2 would constitute between 30 to 80% of the total dynamical mass (compared to 4C 60.07 [27]), if the standard CO/H_2 conversion ratio was adopted. The application of this conversion ratio is however doubtful, and it is possible that the involved H_2 masses are 3–4 times lower[37].

Extremely red objects (EROs, $R - K > 5$) have been identified in searches for primaeval galaxies[15]. A non-negligible fraction ($\sim 10\%$) of the submm sources could be EROs[39]. A proto-typical ERO at $z=1.44$ [11] has been detected in submm continuum[9] and in CO lines[1].

The current CO line detections at high z are summarized in Table 1, and the molecular masses as a function of redshift are displayed in Fig.2. It is clear from this figure that only the very strongest objects are detected, and in particular the gravitationally amplified ones. This situation will rapidly

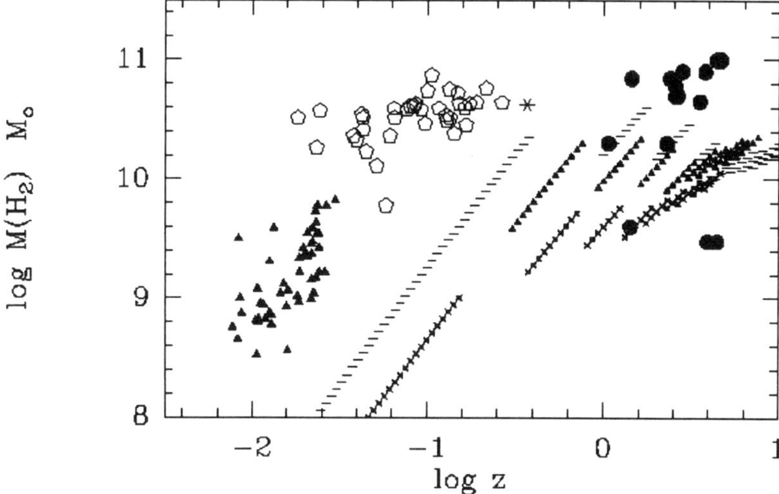

Figure 1. H_2 masses for the CO-detected objects at high redshift (full dots), compared to an ultra-luminous-IR galaxy sample [37] (open pentagons), to a sample of galaxies in the Coma supercluster[8] (filled triangles), and to the quasar 3C48, marked as a star[31,40]. The lines delineated by various symbols indicate the 1σ detection limit at the IRAM-30m telescope of $S(CO) = 1.0$ Jy km/s with the 3 mm receiver (hyphens), 2.0 Jy km/s with the 2 mm (triangles) and 1.3 mm (crosses) receivers, rms that we can reach in a 6 hour integration. Note the absence of detected objects between $z = 0.36 - 1$ in redshift. Due to gravitational amplification the points at high z can be detected well below the sensitivity limit.

change with the new millimeter instruments planned over the world (the NRAO Green-Bank-100m, the UMASS/INAOE LMT-50m, the ALMA (Europe/USA) and the LMSA (Japan) millimetre interferometers). It is therefore interesting to predict with simple models the detection capabilities, as a function of redshift, metallicity or physical conditions in the high-z objects.

3   Perspectives with Future mm Instruments

Sources have been detected in the submillimeter domain at high redshifts because, around 1mm, starbursts have a similar or larger apparent flux at $z = 3$ than at $z = 1$. The line emission does not have such a strong negative k-correction, since in the low frequency domain, the flux of the successive lines increases roughly as ν^2 (optically thick domain), instead of ν^4 for the contin-

Table 1. CO data for high redshift objects

Source	z	CO line	S mJy	ΔV km/s	MH$_2$ 10^{10} M$_\odot$	Ref
SMM 02399-0134	1.062	2-1	3	500	2*	1
0957+561	1.414	2-1	3	440	0.4*	2
HR10	1.439	2-1	4	400	7	3
F10214+4724	2.285	3-2	18	230	2*	4
53W002	2.394	3-2	3	540	7	5
H 1413+117	2.558	3-2	23	330	2-6 *	6
SMM 14011+0252	2.565	3-2	13	200	5*	7
MG 0414+0534	2.639	3-2	4	580	5*	8
SMM 02399-0136	2.808	3-2	4	710	8*	9
6C1909+722	3.532	4-3	2	530	4.5	10
4C60.07	3.791	4-3	1.7	1000	8	10
APM 08279+5255	3.911	4-3	6	400	0.3*	11
BR 1335-0414	4.407	5-4	7	420	10	12
BR 0952-0115	4.434	5-4	4	230	0.3*	13
BR 1202-0725	4.690	5-4	8	320	10	14

* corrected for magnification, when estimated
Masses have been rescaled to $H_0 = 75$km/s/Mpc. When multiple images are
resolved, the flux corresponds to their sum
(1) Kneib et al. (2000); (2) Planesas et al. (1999) (3) Andreani et al. (2000);
(4) Solomon et al. (1992), Downes et al. (1995); (5) Scoville et al. (1997b); (6)
Barvainis et al. (1994); (7) Frayer et al. (1999); (8) Barvainis et al. (1998);
(9) Frayer et al. (1998); (10) Papadopoulos et al. (2000); (11) Downes et
al. (1999); (12) Guilloteau et al. (1997); (13) Guilloteau et al. (1999); (14)
Omont et al. (1996)

uum. Nevertheless the line emission is essential to study the nature of the
object (the AGN-starburst connection for instance), and deduce their physical conditions (kinematics, abundances, excitation). Given the gas and dust
temperatures, the maximum flux is always reached at much lower frequencies
than in the continuum, since the lines always reflect the energy difference between two levels. This is an advantage, given the largest atmospheric opacity
at high frequencies.

3.1 Predicted Line and Continuum Fluxes

To model high-redshift starburst objects, let us extrapolate the properties of more local ones: the active region is generally confined to a compact nuclear disk, sub-kpc in size[32,35,37]. The gas is much denser here than the average over a normal galaxy, of the order of 10^4 cm^{-3}, with clumps at least of 10^6 cm^{-3} to explain the data on high density tracers (HCN, CS). The ISM maybe modelled by two density and temperature components[10], at 30 and 90K. The total molecular mass considered will be 6×10^{10} M$_\odot$ and the average column density N(H$_2$) of 10^{24} cm^{-2}, typical of the Orion cloud center.

Going towards high redshift ($z > 9$), the temperature of the cosmic background T_{bg} becomes of the same order as the interstellar dust temperature, and the excitation of the gas by the background radiation competes with that of gas collisions. It might then appear easier to detect the lines[34], but this is not the case when every effect is taken into account. To have an idea of the increase of the dust temperature with z, the simplest assumption is to consider the same heating power due to the starburst. At a stationary state, the dust must then radiate the same energy in the far-infrared that it receives from the stars, and this is proportional to the quantity $T_{dust}^6 - T_{bg}^6$, if the dust is optically thin, and its opacity varies as ν^β, with $\beta = 2$. Keeping this quantity constant means that the energy re-radiated by the dust, proportional to T_{dust}^6, is always equal to the energy it received from the cosmic background, proportional to T_{bg}^6, plus the constant energy flux coming from the stars. Since β can also be equal to 1 or 1.5, or the dust be optically thick, we have also considered the possibility of keeping $T_{dust}^4 - T_{bg}^4$ constant; this does not change fundamentally the results.

Computing the populations of the CO rotational levels with an LVG code, and in the case of the two component models described earlier, the predictions of the line and continuum intensities as a function of redshift and frequencies are plotted in Fig.2.

3.2 Source Counts

To predict the number of sources that will become available with the future sensitivity, let us adopt a simple model of starburst formation, in the frame of the hierarchical theory of galaxy formation. The cosmology adopted here is an Einstein-de Sitter model, $\Omega = 1$, with no cosmological constant, and $H_0 = 75 \mathrm{kms}^{-1}\mathrm{Mpc}^{-1}$, $q_0 = 0.5$. The number of mergers as a function of redshifts can be easily computed through the Press-Schechter formalism[29], assuming self-similarity for the probability of dark halos merging, and an ef-

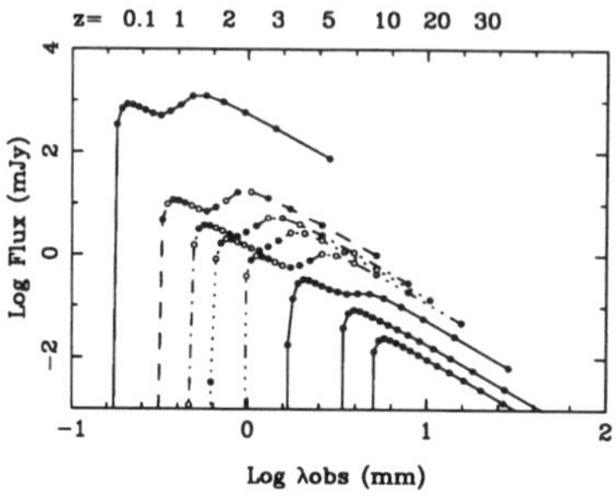
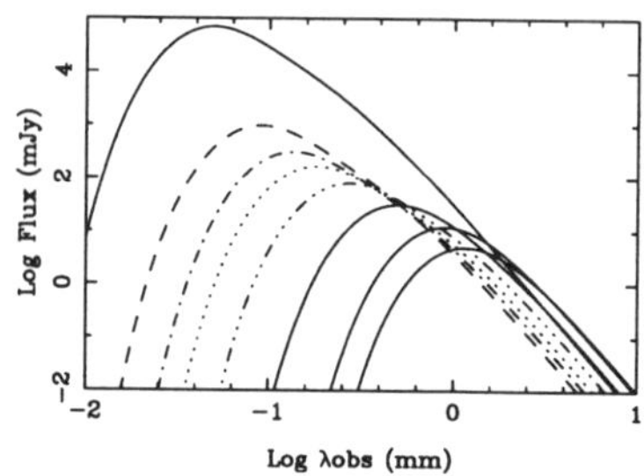

Figure 2. Expected flux for the two-component cloud model, for various redshifts $z = 0.1$, 1, 2, 3, 5, 10, 20, 30, and $q_0 = 0.5$. **Left** are the CO lines, materialised each by a circle (they are joined by a line only to guide the eye). **Right** is the continuum emission from dust. It has been assumed here that $T_{dust}^6 - T_{bg}^6$ is conserved (from Combes et al. 1999).

ficiency of mergers in terms of star-formation peaking at $z \sim 2$ (*i.e.* Blain & Longair 1993). Also, the integration of the flux of all sources, over all redshifts, should agree with the cosmic infrared background detected by COBE. These assumptions and restrictions reduce the number of free parameters in the model. To fit the source counts, however, another parameter must be introduced which measures the rate of energy released in a merger (or the life-time of the event). This rate must increase strongly with redshift[5]. Once the counts are made compatible with the submm observations, the model determines the contribution of sources in different redshift intervals to the present counts. It is interesting to note that the intermediate redshifts dominate the continuum source counts ($2 < z < 5$) if we allow the star formation to begin before $z = 6$. At higher dust temperatures the counts are dominated by the highest redshifts ($z > 5$).

Once these fits are obtained for the continuum sources, it is possible to derive also the counts for CO line emission from the same sources. The spectral energy distribution is now obtained with a comb-like function, representing the rotational ladder, convolved with a Planck distribution of temperature equal to the dust temperature, assuming the lines are optically thick. The frequency filling factor is then proportional to the rotational number, and therefore to the redshift, for a given observed frequency. The width of the lines have been assumed to be $300 \mathrm{kms}^{-1}$. The derived source numbers are shown in Fig.3. Note how they are dominated by the high redshift sources. It is not useful to observe at λ below $1\,\mathrm{mm}$ for high-z protogalaxies, but instead to move the observing wavelength towards $\lambda = 1\,\mathrm{cm}$.

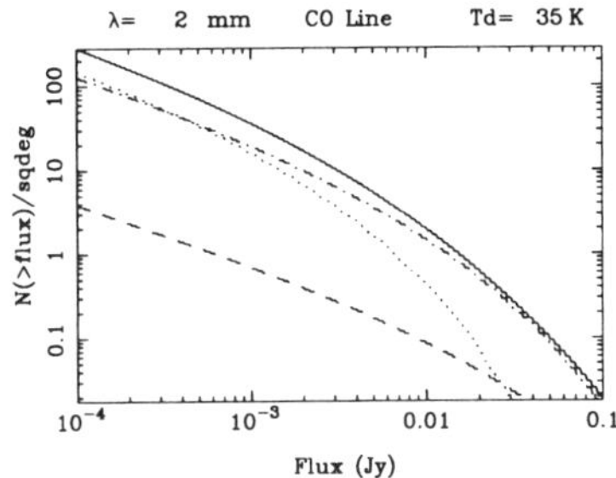
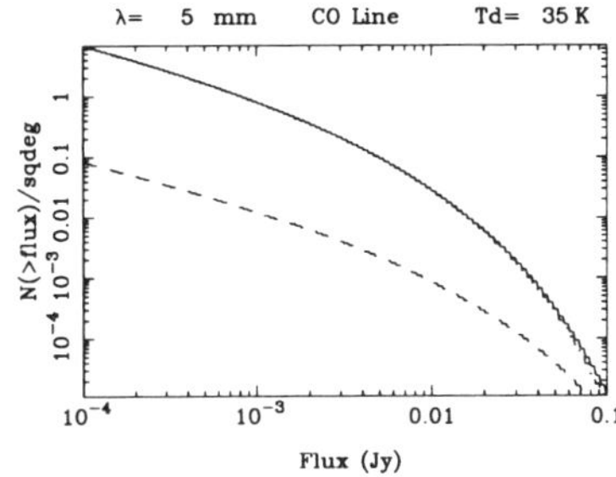

Figure 3. *Left* Source counts for the CO lines at an observed frequency of 2 mm, assuming optically thick gas at $T_{ex} = 35$ K. The solid line represents the total counts. The dashed line, dot-dashed and dotted lines show the counts from sources at the lowest redshifts ($z < 2$), intermediate redshifts ($2 < z < 5$) and highest redshifts ($z > 5$) respectively. *Right* Same for $\lambda = 5$ mm.

4　Conclusion

The search for CO lines in high redshift galaxies has only just begun. There will be a real breakthrough in the next decade, when the sensitivity is increased by more than a factor 10 with the next generation of millimeter instruments. The detection of CO lines is fundamental to increased understanding of star formation and its efficiency at high z, to measure molecular and total gas masses, dynamical masses and to differentiate AGN processes from from starburst activity in distant galaxies.

Acknowledgements

I am very grateful to James Lowenthal and David Hughes for the organisation of such an interesting meeting, and to UMass for their financial help.

References

1. Andreani, P., Cimatti, A., Loinard, L. *et al*, *A&A* **354**, L1 (2000).
2. Barger, A.J., Cowie, L.L., Sanders, D.B., *ApJ* **518**, L5 (1999).
3. Barvainis R., Tacconi L., Antonucci R. *et al*, *Nature* **371**, 586 (1994).
4. Barvainis R., Alloin D., Guilloteau S. *et al*, *ApJ* **492**, L13 (1998).
5. Blain A.W., Jameson A., Smail I. *et al*, *MNRAS* **309**, 715 (1999).
6. Blain A.W., Longair M.S., *MNRAS* **264**, 509 (1993).
7. Brown R., Vanden Bout P., *ApJ* **397**, L19 (1992).

8. Casoli F., Dickey J., Kazes I. *et al*, *A&AS* **116**, 193 (1996).

9. Cimatti A., Andreani P., Röttgering H. *et al*, *Nature* **392**, 895 (1998).

10. Combes F., Maoli R., Omont A., *A&A* **345**, 369 (1999).

11. Dey, A., Graham, J.R., Ivison, R.J. *et al*, *ApJ* **519**, 610 (1999).

12. Downes D., Neri R., Wiklind T. *et al*, *ApJ* **513**, L1 (1999).

13. Downes D., Solomon P.M., Radford S.J.E., *ApJ* **453**, L65 (1995).

14. Elbaz D., Arnaud M., Casse M. *et al*, *A&A* **265**, L29 (1992).

15. Elston R., Rieke G.H., Rieke M.J., *ApJ* **331**, L77 (1988).

16. Frayer D.T., Ivison R.J., Scoville N.Z. *et al*, *ApJ* **506**, L7 (1998).

17. Frayer D.T., Ivison R.J., Scoville N.Z. *et al*, *ApJ* **514**, L13 (1999).

18. Guiderdoni B. *et al*, *Nature* **390**, 257 (1997).

19. Guilloteau S., Omont A., Cox P. *et al*, *A&A* **349**, 363 (1999).

20. Guilloteau S., Omont A., McMahon R.G. *et al*, *A&A* **328**, L1 (1997).

21. Hauser, M. *et al*, *ApJ* **508**, 25 (1998).

22. Hughes D.H., Serjeant S., Dunlop J. *et al*, *Nature* **394**, 241 (1998).

23. Kneib J-P. *et al*, *A&A*, in preparation (2000).

24. Madau, P. *et al*, *MNRAS* **283**, 1388 (1996).

25. Ohta K., Yamada T., Nakanishi K. *et al*, *Nature* **382**, 426 (1996).

26. Omont A., Petitjean P., Guilloteau S. *et al*, *Nature* **382**, 428 (1996).

27. Papadopoulos P.P., Röttgering H.J.A. *et al*, *ApJ* **528**, 626 (2000).

28. Planesas, P. *et al*, *Science* **286**, 2493 (1999).

29. Press W.H., Schechter P., *ApJ* **187**, 425 (1974).

30. Puget, J.-L., Abergel, A., Bernard, J.-P. *et al*, *A&A* **308**, L5 (1996).

31. Scoville N.Z., Padin S., Sanders D.B. *et al*, *ApJ* **415**, L75 (1993).

32. Scoville N.Z., Yun M.S., Bryant P.M., *ApJ* **484**, 702 (1997a).

33. Scoville N.Z., Yun M.S., Windhorst R.A. *et al*, *ApJ* **485**, L21 (1997b).

34. Silk J., Spaans M., *ApJ* **488**, L79 (1997).

35. Solomon P.M., Radford S.J.E., Downes D., *ApJ* **348**, L53 (1990).

36. Solomon P.M., Downes D., Radford S.J.E., *Nature* **356**, 318 (1992).

37. Solomon P.M. *et al*, *ApJ* **478**, 144 (1997).

38. Smail, I., Ivison, R.J., Blain, A.W., *ApJ* **490**, L5 (1997).

39. Smail, I., Ivison, R.J., Kneib, J-P. *et al*, *MNRAS* **308**, 1061 (1999).

40. Wink J.E., Guilloteau S., Wilson T.L., *A&A* **322**, 427 (1997).

CO and Near-Infrared Observations of High-Redshift Submillimeter Galaxies

D. T. Frayer

California Institute of Technology
Astronomy Dept. 105-24
Pasadena, CA 91125
E-mail: dtf@astro.caltech.edu

I discuss our ongoing Owens Valley Millimeter Array observations and Keck near-infrared wavelength observations of the high-redshift sub-mm population of galaxies. These observations are important for our understanding of the distant universe since the sub-mm population accounts for a large fraction of the extragalactic background at mm/sub-mm wavelengths and contributes significantly to the total amount of star-formation and AGN activity at high redshift. The CO data suggest that the sub-mm galaxies are analogous to the gas-rich ultraluminous systems found in the local universe. Initial near-infrared data show that many of the sub-mm galaxies are faint-red sources which are undetected at ultraviolet/optical wavelengths. These results highlight the importance that future sensitive mm-wavelength instruments, such as the LMT and ALMA, will have on our understanding of the early evolution and formation of galaxies.

1 Introduction

The discovery of an ultraluminous population of high-redshift galaxies with deep submillimeter surveys has revolutionized our understanding of the distant universe.[25,2,19,12,6] The current data show that the sub-mm population has a mixture of AGN and starburst characteristics with properties which are roughly consistent with the local population of ultraluminous ($L > 10^{12} L_\odot$) infrared galaxies (ULIGs). The relative importance of AGN and starburst activity in powering the high luminosities of the sub-mm population is still an open question, but the growing consensus is that the majority of the luminosity of the population is powered by star formation.[7] The early CO and X-ray data on the sub-mm population support the starburst nature of the population by showing the presence of sufficient molecular gas to fuel the star-formation activity[15,14] and the lack of expected X-ray emission if mostly dominated by AGN.[13,18,24,1]

Although the redshift distribution of the sub-mm population is still uncertain, the majority of the sub-mm galaxies are believed to be at high redshifts ($z \gtrsim 2$) based on their radio[9,28] and near-infrared[27] data. The early redshift distributions based on optical imaging and spectroscopy suggested somewhat lower redshifts,[26,3,23] but several of the original candidate optical counter-parts

Table 1: Brightest Sources in SCUBA Cluster Lens Survey

Galaxy	$S(850\mu m)^3$ (mJy)	Redshift	$K-$mag	Notes
SMM J02399$-$0136	25.4	2.808 [15]	19.1 [21]	CO[15]
SMM J00266$+$1708	18.6	> 2.0 [28]	22.5 [16]	
SMM J09429$+$4658	17.2	> 3.9 [28]	19.4 [27]	ERO-H5 [27]
SMM J14009$+$0252	14.5	> 0.7 [28]	21.0 [20]	J5 [20]
SMM J14011$+$0252	12.3	2.565 [14]	17.8 [20]	CO[14]
SMM J02399$-$0134	11.0	1.062 [29]	16.3 [29]	CO[22]
SMM J22471$-$0206	9.2	> 1.8 [28]	?	
SMM J02400$-$0134	7.6	> 2.4 [28]	?	
SMM J04431$+$0210	7.2	> 1.6 [28]	19.1 [27]	ERO-N4 [27]

have turned out to be incorrect. Despite their ultra-high luminosities, many sub-mm galaxies are undetected at ultraviolet/optical wavelengths due to extinction by dust. For these highly obscured galaxies, follow-up radio[28] and/or mm interferometric observations[10,16,4,17] are required in order to uncover the proper counter-part.

In order to understand the nature of the sub-mm population, we have been carrying out multi-wavelength observations of individual systems in the SCUBA Cluster Lens Survey.[26] This survey represents sensitive sub-mm mapping of seven massive, lensing clusters which uncovered 15 background sub-mm sources. The advantages of this sample are that the amplification of the background sources allows for deeper source frame observations and that lensing by cluster potentials does not suffer from differential lensing. We have concentrated our efforts on the nine background galaxies detected at the highest signal–to–noise (Table 1). Only three sources have spectroscopic redshifts, and the redshift lower limits shown in Table 1 are based on their sub-mm/radio flux ratios.[28]

2   CO Results

At OVRO we have conclusively detected CO emission from two sub-mm systems, SMM J02399$-$0136 at $z = 2.8$ (SMM J02399) and SMM J14011$+$0252 at $z = 2.6$ (SMM J14011) (Fig. 1). A third system SMM J02399$-$0134, which is associated with a ring-galaxy containing a Seyfert nucleus at $z = 1.06$,[29] has recently been detected in CO at the PdB.[22] We have also tentatively confirmed the PdB detection at OVRO. To date, these three galaxies are the only sub-mm sources with known redshifts, and it is promising that all three have already been detected in CO. The early CO results suggest that the sub-mm population contains massive reservoirs of molecular gas and are among the most CO luminous galaxies in the universe.

The strongest sub-mm source, SMM J02399, shows an AGN component in its optical spectrum,[21] while SMM J14011 shows only evidence for starburst activity at optical/NIR wavelengths.[20] Although the optical characteristics of these two galaxies are vastly different, their radio, sub-mm, and CO properties are fairly similar and are consistent with a high level of star formation activity (SFRs of a few$\times 10^2\,M_\odot\,yr^{-1}$ to more than $10^3\,M_\odot\,yr^{-1}$, depending on the IMF and AGN contamination). After correcting for lensing, we derive CO luminosities of $3\text{--}4\times 10^{10}\,K\,km\,s^{-1}\,pc^2$ ($H_0 = 50\,km\,s^{-1}\,Mpc^{-1}$; $q_0 = 1/2$) in these two systems. These CO luminosities correspond to molecular gas masses of about $5 \times 10^{10}\text{---}2 \times 10^{11}\,M_\odot$, depending on the exact value of the CO to H_2 conversion factor. Both SMM J02399 and SMM J14011 appear to be associated with a merger event. Given that mergers of gas-rich galaxies at low-redshift result in massive starbursts, we expect star-formation to be an important component for powering the far-infrared luminosities in both of these systems. In fact, the large molecular gas masses of SMM J02399 and SMM J14011 are sufficient to form the stars of an entire L^* galaxy, which suggests that the sub-mm population may represent the formative phase of massive galaxies.

SMM J02399 is unresolved in CO, while SMM J14011 is extended over a large spatial scale in its source frame ($\gtrsim 10\,kpc$). Figure 1 contains low resolution OVRO data which showed tentative evidence for extended CO emission in the north-south direction in SMM J14011. The extended morphology of the molecular gas in SMM J14011 has been recently confirmed with higher-resolution data from OVRO and BIMA. These results may suggest that SMM J14011 is in an early stage of its merger event, unlike the majority of ULIGs in the local universe whose CO emission is mostly contained within the central kpc.[11,8] It is currently unknown what fraction of the sub-mm sources are compact or are extended over large spatial scales as is SMM J14011. If the progenitors of the sub-mm systems are more gas rich than those of local ULIGs, we could expect the sub-mm sources to have larger gas fractions and to be more extended than their low-redshift analogs. A large sample of sub-mm sources need to be observed in CO before statistical comparisons could be made between the CO properties of local ULIGs and the high-redshift sub-mm galaxies.

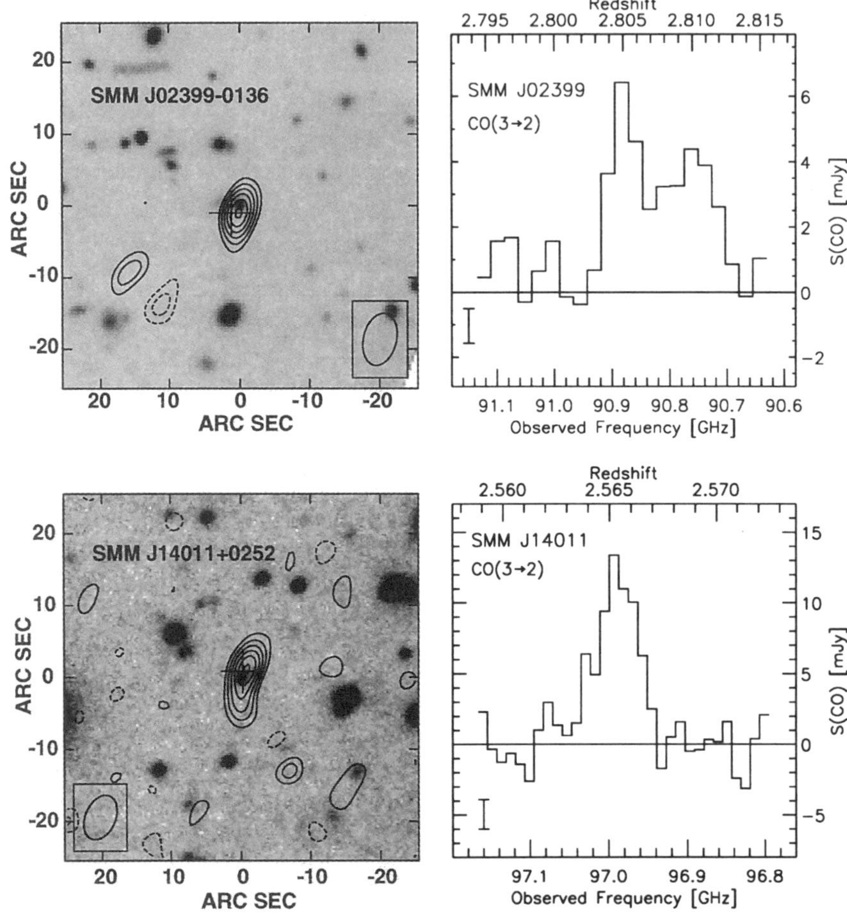

Figure 1: OVRO CO(3-2) detections for SMM J02399−0136 (top) and SMM J14011+0252 (bottom). The grey-scale images at the left are optical images while the contours represent the integrated CO maps for the sub-mm galaxies. The crosses represent the positions of the SCUBA detection. The corresponding CO(3-2) spectra are shown at the right. These data were originally published in Frayer et al. (1998, 1999).

3 Near-Infrared Results

Many sub-mm galaxies are too obscured by dust to be detected at ultraviolet/optical wavelengths. At least four of the nine sources in our sample which were undetected at optical wavelengths ($I > 25$–26) have faint near-infrared, K-band counter-parts. Two of these are bright enough in K-band ($K = 19.1$, 19.6 mag) to be classified as extremely red objects (EROs).[27] An additional faint ($K = 21$) galaxy was found associated with a relatively bright (0.5 mJy) radio counter-part.[20] The fourth and faintest galaxy with a near-infrared counter-part is SMM J00266+1708.

The sub-mm position of SMM J00266+1708 is located between three optically visible galaxies.[26] We imaged the field at OVRO at 1.3 mm and determined its position to be offset from all three optical sources. Deep, follow-up near-infrared observations with NIRC on Keck uncovered a new faint galaxy at $K = 22.5$ located at the position of the 1.3 mm source.[16] Although SMM J00266+1708 is the second brightest sub-mm source in the SCUBA Cluster Lens Survey, it is currently the faintest known near-infrared counter-part of a sub-mm galaxy discovered to date (Fig. 2).

Only two of the nine sources in the sample still require deep K-band imaging and currently have uncertain counter-parts. The galaxy SMM J02400−0134 has no optically detected galaxies near the sub-mm position. For SMM J22471-0206 there are several optical galaxies which could be the sub-mm counter-part, but given previous results it will be interesting to test whether or not any new candidate galaxies are uncovered with deep K-band imaging. Depending on the results for these last two unknown systems, the current data suggest that approximately 40%–70% (4/9–6/9) of the sub-mm population as a whole have faint near–infrared counter-parts that are undetected at optical wavelengths. Only 30%–40% (3/9–4/9) of the sample have optical counterparts ($I <26$–27, correcting for source lensing).

The K-band magnitudes of the sub-mm counter-parts in the SCUBA Cluster Lens Survey range over 6 magnitudes which reflects the wide diversity of colors and redshifts for the population (Table 1). The magnitudes listed in Table 1 have not been corrected for lensing; unlensed sources would be about a magnitude fainter on average. Currently, only the three brightest optical (I-band) sources have spectroscopic redshifts. The other sources are much fainter and would require near-infrared spectroscopy to obtain redshifts, which will be challenging even with 8m/10m class ground base telescopes. Our early spectroscopic results with NIRSPEC on Keck suggest that lines may be detectable in the brighter sources ($K \lesssim 20$), while many of the fainter sub-mm sources ($K \gtrsim 22$) may have to wait for the *Next Generation Space Telescope*.

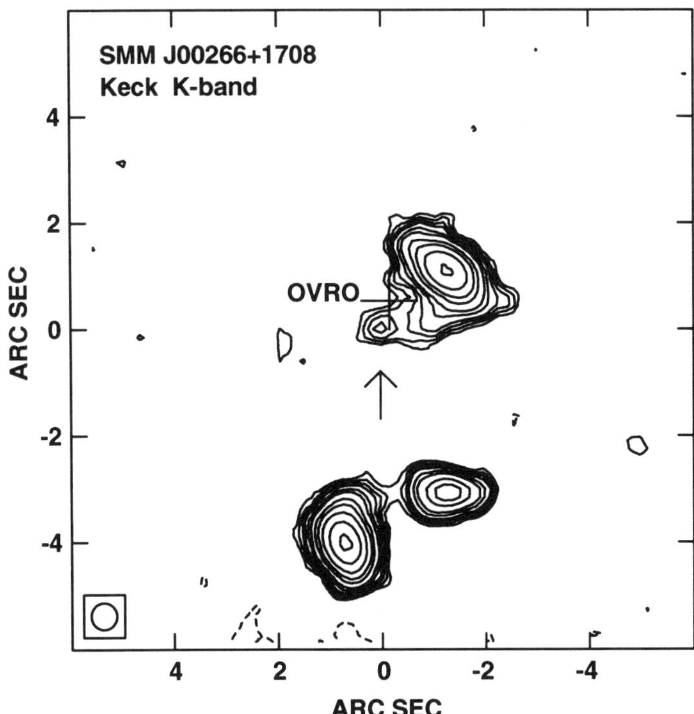

Figure 2: The Keck K-band (2.2μm) image of SMM J00266+1708. The near-infrared data were taken after determining the accurate position of the sub-mm source with OVRO 1.3 mm imaging. The position of the 1.3 mm source is shown by the cross labeled "OVRO". The three bright sources in the field are foreground galaxies previously observed at optical wavelengths. The arrow points to the new galaxy not detected in the optical ($I > 26$) thought to be the counter-part of SMM J00266+1708. The rms of the image is 24.8 mag/sq-arcsec (0.04μJy/beam), and the contours are $1\sigma \times (-3, 3, 4, 5, 6, 8, 10, 15, 20, 30, 50, 80)$. The seeing disk (beam) of the near-infrared data is shown in the lower left ($0.5 \times 0.5''$). These data were originally published in Frayer et al. (2000).

4 Conclusions

Most sub-mm sources are too red and/or faint to be detected at optical wavelengths. There is very little overlap, if any, between the ultraluminous sub-mm population and the less luminous, optically-selected Lyman Break population of galaxies. This highlights the importance of radio, millimeter, and near-infrared observations of the sub-mm population for our general understanding of the evolution and formation of galaxies.

Potentially, we do not need to wait for future optical/near-infrared space-based missions to obtain redshifts for the bulk of the sub-mm population. Redshifts could be determined directly from the CO lines themselves with planned ground-based millimeter telescopes, such as the LMT and ALMA. Both ALMA and the LMT will have sufficient sensitivities and broad-bandwidth spectrometer capabilities to make large CO redshift surveys practical.[5] The proposed 30 GHz spectrometer for the LMT would be an excellent redshift machine for the sub-mm population of galaxies.

The two best studied sub-mm galaxies (SMM J02399−0136 and SMM J14011+0252) share many of the same properties of the local population of ULIGs, such as high infrared luminosities, the association with mergers, massive molecular gas reservoirs, comparable CO line widths, and similar IR/radio and IR/CO luminosity ratios. Future CO observations of large samples of ultraluminous galaxies with ALMA and the LMT will enable us to study the evolution of the molecular gas properties as a function of redshift, which will be crucial for understanding the star-formation history of the universe.

Acknowledgments

I thank the work of my collaborators Nick Scoville, Rob Ivison, Ian Smail, Andrew Blain, Aaron Evans, Min Yun, and Jean-Paul Kneib. I appreciate the efforts of the OVRO and Keck staff who have made these observations a success. I acknowledge support from NSF grant AST 9981546 made to the OVRO Millimeter Array, which is operated by the California Institute of Technology. I thank the organizers at the University of Massachusetts and INAOE for planning the conference and providing support to attend the conference.

1. Barger, A. J., Cowie, L. L., Mushotzky, R. F., & Richards, E. A. 2000, AJ submitted (astro-ph/0007175)
2. Barger, A. J., Cowie, L. L., Sanders, D. B., Fulton, E., Taniguchi, Y., Sato, Y., Kawara, K., & Okuda, H. 1998, Nature, 394, 248
3. Barger, A. J., Cowie, L. L., Smail, I., Ivison, R. J., Blain, A. W., & Kneib, J.-P. 1999, AJ, 117, 2656

4. Bertoldi, F., et al. 2000, A&A, in press (astro-ph/0006094)

5. Blain, A. W., Frayer, D. T., Bock, J. J., & Scoville, N. Z. 2000, MNRAS, 313, 559

6. Blain, A. W., Kneib, J.-P., Ivison, R. J., & Smail, I. 1999a, ApJ, 512, L87

7. Blain, A. W., Smail, I., Ivison, R. J., & Kneib, J.-P. 1999b, MNRAS, 302, 632

8. Bryant, P. M., & Scoville, N. Z. 1999, AJ, 117, 2632

9. Carilli, C. L., & Yun, M. S. 1999, ApJ, 513, L13

10. Downes, D., et al. 1999, A&A, 347, 809

11. Downes, D., & Solomon, P. M. 1998, ApJ, 507, 615

12. Eales, S., Lilly, S., Gear, W., Dunne, L., Bond, J. R., Hammer, F., Le Fèvre, O., & Crampton, D. 1999, ApJ, 515, 518

13. Fabian, A. C., et al. 2000, MNRAS, 315, L8

14. Frayer, D. T., et al. 1999, ApJ, 514, L13

15. Frayer, D. T., Ivison, R., J., Scoville, N. Z., Yun, M., Evans, A. S., Smail, I., Blain, A. W., &, Kneib, J.-P. 1998, ApJ, 506, L7

16. Frayer, D. T., Smail, I., Ivison, R. J., & Scoville, N. Z. 2000, AJ, in press (astro-ph/0005239)

17. Gear, W. K., Lilly, S. J., Stevens, J. A., Clements, D. L., Webb, T. M., Eales, S. A., & Dunne, L. 2000, MNRAS, in press (astro-ph/0007054)

18. Hornschemeier, A. E., et al. 2000, ApJ, in press (astro-ph/0004260)

19. Hughes, D., et al. 1998, Nature, 394, 241

20. Ivison, R. J., Smail, I., Barger, A. J., Kneib, J.-P., Blain, A. W., Owen, F. N., Kerr, T. H., & Cowie, L. L. 2000, MNRAS, 315, 209

21. Ivison, R. J., Smail, I., Le Borgne, J.-F., Blain, A. W., Kneib, J.-P., Bézecourt, J., Kerr, T. H., & Davies, J. K. 1998, MNRAS, 298, 583

22. Kneib, J.-P., et al. 2000, in preparation

23. Lilly, S. J., Eales, S., A., Gear, W. K. P., Hammer, F., Le Fèvre, O., Crampton, D., Bond, J. R., & Dunne, L. 1999, ApJ, 518, 641

24. Severgnini, P., et al. 2000, A&A, in press (astro-ph/000623)

25. Smail, I., Ivison, R. J., & Blain, A. W. 1997, ApJ, 490, L5

26. Smail, I., Ivison, R. J., Blain, A. W., & Kneib, J.-P. 1998, ApJ, 507, L21

27. Smail, I., Ivison, R. J., Kneib, J.-P., Cowie, L. L., Blain, A. W., Barger, A. J., Owen, F. N., & Morrison, G. 1999, MNRAS, 308, 1061

28. Smail, I., Ivison, R. J., Owen, F. N., Blain, A. W., & Kneib, J.-P. 2000, ApJ, 528, 612

29. Soucail, G., Kneib, J. P., Bézecourt, J., Metcalfe, L., Altieri, B., Le Borgne, J. F. 1999, A&A, 343, L70

6. Clustering

THE FIR-RADIO CORRELATION IN NEARBY CLUSTERS: IMPLICATIONS FOR THE RADIO-TO-SUBMM INDEX REDSHIFT ESTIMATOR

NEAL A. MILLER

National Radio Astronomy Observatory, 1003 Lopezville Road, Socorro, NM, 87801
and New Mexico State University
E-mail: nmiller@nrao.edu

FRAZER N. OWEN

National Radio Astronomy Observatory, 1003 Lopezville Road, Socorro, NM, 87801
E-mail: fowen@nrao.edu

The radio-to-submillimeter spectral index as a redshift indicator is a powerful technique for submm sources[1,2]. As it is predicated upon the far-infrared-to-radio correlation (FIR-radio correlation), a comprehensive evaluation of this correlation in the nearby universe elucidates the potential uncertainties inherent to the technique. Using a comprehensive data set covering 17 nearby Abell clusters, we assess the FIR-radio correlation of galaxies from the cluster cores out well past the Abell radius. For the majority of galaxies, the FIR-radio correlation is confirmed. This includes AGN, though such objects exhibit increased radio emission relative to that expected from normal star forming objects. A slight excess of radio over-luminous spirals in the cores of clusters is also identified, and high resolution radio imaging confirms that the potential AGN contribution to their radio flux is insignificant. The relative numbers of 'normal' galaxies and outliers to the FIR-radio correlation in nearby clusters may be used to assess the redshift errors from applying the radio-submm spectral index technique to higher redshift galaxies. The power of the technique in statistical studies is confirmed.

1 Introduction

Radio observations play a key role in studies of submm sources. The coarse resolution of current submm detectors often implies that the submm emission could arise from any of several optical sources, which is further complicated by such sources often being very faint and red. The resolution of the VLA ordinarily provides unambiguous identification of the true optical counterpart, and as noted in Carilli & Yun[1] the spectral index determined from the radio and submm fluxes provides an estimate of the redshift of the source.

The underlying cause of these benefits to radio observation is the spectral energy distribution (SED) of galaxies engaged in star formation. The SED of a zero-redshift galaxy is dominated by synchrotron emission at radio wavelengths and thermal emission at infrared wavelenths, and is smooth over the frequency range $\sim 10^9 - \sim 10^{13}$Hz. The spectral index turns over around

128

10^{11}Hz, with synchrotron describing the lower frequencies ($\alpha \sim -0.8$, where $S_\nu \propto \nu^\alpha$) and the modified blackbody of the thermal peak describing the higher frequencies ($\alpha \sim 3.0 - 3.5$). The FIR $60\mu m$ and $100\mu m$ fluxes occur very near the thermal peak, and observations of many nearby galaxies fix the ratio of this peak to the radio emission (see Condon 1992 for a review[3]). Since $850\mu m$ occurs near the base of the thermal peak at zero redshift and the spectral indices are so smooth over large frequency ranges, submm sources at redshifts of $\sim 1 - 6$ are relatively strong and knowledge of the SED allows for an accurate photometric redshift to be obtained. Carilli & Yun represent this as

$$\alpha_{1.4}^{350} = -0.24 - [0.42 \times (\alpha_{radio} - \alpha_{submm}) \times \log(1 + z)] \qquad (1)$$

where $\alpha_{1.4}^{350}$ represents the spectral index between 1.4GHz (20cm) and 350GHz ($850\mu m$).

The assumptions inherent in this approach lead to several sources of potential error. The spectral index for the radio may differ (free-free absorption, emission) as may the thermal spectral index (dust emissivity). Inverse Compton losses to the CMB may affect the radio fluxes at high redshift, though at such redshifts the $850\mu m$ bandpass has shifted over the thermal peak anyway. Variations in dust temperature are an additional potential source of error[4]. Perhaps the greatest source of error is galaxies which lie off the FIR-radio correlation — AGN or unusual star forming galaxies — which is the focus of this study.

2 Data

The sample consists of 329 radio galaxies drawn from 17 nearby clusters for which $z < 0.033$[5]. The galaxies are identified from the cluster cores out to $3h_{75}^{-1}$Mpc, with optical spectroscopy and/or public redshifts confirming their cluster membership. Their radio fluxes are taken from the NVSS[6] and their FIR fluxes from the IRAS catalogs or cross scans of the IRAS data. Optical spectroscopy allowed for characterization as star formation or AGN for many of the galaxies. The FIR-radio correlation was parametrized by the statistic q, which is the logarithmic ratio of the FIR flux to the radio flux[7]. The average value for q in star forming galaxies is 2.3, with a dispersion of only 0.2.

Prior studies have suggested that cluster star forming galaxies are often over-luminous in the radio[8,9], usually attributed to ram pressure compression of their B fields as they plow through the intracluster gas. The large radial extent of the database allows for investigation of this trend without the bias caused by differing selection of field and cluster samples, and the spectroscopy minimizes foreground/background contamination and removes AGN from the star forming sample. In addition, 8.46GHz VLA A-array observations of a subset of the radio over-luminous star forming galaxies provide the resolution (0.25″; unresolved sources have size limits of $\sim$ 25pc) to determine whether an AGN contributes to the net radio flux of such objects.

3 Results

The FIR-radio correlation was found to hold strongly in the nearby clusters. While a population of radio over-luminous star forming galaxies was identified, such objects are rare and represent only a few percent of the total population in a rich cluster and are generally absent from poorer clusters. Such galaxies are preferentially located in the cluster cores, though detection of this effect results mainly from the large sample size. It is likely that their frequency has been overestimated in prior studies due to the inclusion of AGN. Furthermore, they are generally over-luminous in the radio by only a factor of 2–3 and not the factor of 5–10 claimed previously[9]. The high resolution radio observations do not detect compact radio emission from these objects, ruling out AGN contamination and constraining the star formation to occur on size scales greater than $\sim$100 pc. The FIR and radio fluxes for AGN (with significant FIR flux) are also strongly correlated. These also display a slight radial dependence, with centrally-located objects more likely to be radio-overluminous. Such objects are between two and five times more radio-loud than the standard FIR-radio correlation would predict.

4 Application

Given these results, it is easy to estimate the errors in redshift that would arise from application of Equation 1 to the nearby cluster populations if they were located at higher redshift. Assuming the same general spectral energy distribution (i.e., $\alpha_{radio} = -0.8$ and $\alpha_{submm} = 3.5$) but with the radio flux boosted by the value suggested by deviations from the FIR-radio correlation, we calculate the redshift error as a function of galaxy redshift for the cases of a galaxy over-luminous in the radio by a factor of two and a factor of five (see Figure 1). The excess radio emission causes an underestimate of the true redshift of the source, in a manner that is strongly dependent on redshift nearby and nearly independent of redshift for $z >\sim 2$ (due to the location of the turnover in the SED from synchrotron to thermal emission). Placing a cluster such as Coma at $z = 2$ (Coma has the highest fraction of 'unusual' objects of any cluster in the nearby sample) gives an idea of the potential error

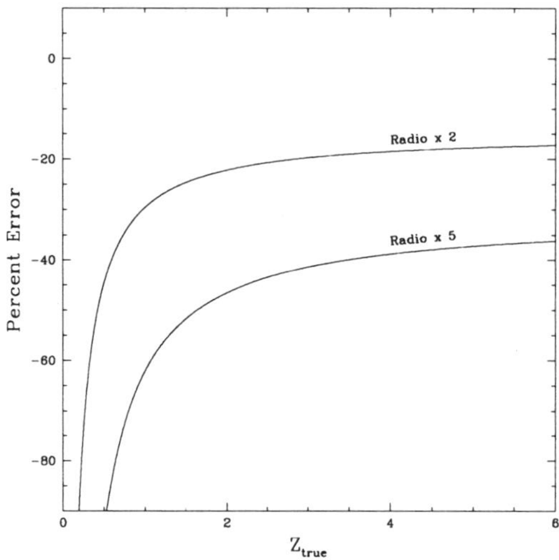

Figure 1. Percent error in redshift estimated via radio-to-submm technique for radio over-luminous sources.

in applying the radio-to-submm technique. Adopting the relative fraction of normal star forming galaxies, radio over-luminous star forming galaxies, and AGN observed in Coma would produce a weighted average redshift only 15% less than the true redshift of the cluster. Thus, conclusions drawn from large samples of submm objects with estimated redshifts are fairly robust.

References

1. Carilli, C.L., and Yun, M.S. 1999, ApJ, 513, L13
2. Carilli, C.L., and Yun, M.S. 2000, ApJ, 530, 618
3. Condon, J.J. 1992, ARA&A, 30, 575
4. Blain, A.W. 1999, MNRAS, 309, 955
5. Miller, N.A., and Owen, F.N. 2000, in preparation
6. Condon, J.J., Cotton, W.D., Griesen, E.W., Yin, Q.F., Perley, R.A., Taylor, G.B., and Broderick, J.J. 1998, AJ, 115, 1693
7. Helou, G., Soifer, B.T., and Rowan-Robinson, M. 1985, ApJ, 298, L7
8. Andersen, V., and Owen, F.N. 1995, AJ, 109, 1582
9. Gavazzi, G., and Boselli, A. 1999, A&A, 343, 93

CLUSTERING IN DEEP (SUBMILLIMETRE) SURVEYS

ENRIQUE GAZTAÑAGA & DAVID H. HUGHES

Instituto Nacional de Astrofísica, Óptica y Electrónica (INAOE),
Luis Enrique Erro 1, Tonantzintla, Cholula, 78840 Puebla, Mexico

Hughes & Gaztañaga (2001, see article in these proceedings) have presented realistic simulations to address key issues confronting existing and forthcoming submm surveys. An important aspect illustrated by the simulations is the effect induced on the counts by the sampling variance of the large-scale galaxy clustering. We find factors of up to $\sim 2-4$ variation (from the mean) in the extracted counts from deep surveys identical in area ($\sim 6\,\mathrm{arcmin}^2$) to the SCUBA surveys of the Hubble Deep Fields (HDF) [4]. Here we present a recipe to model the expected degree of clustering as a function of sample area and redshift.

1 A model for the angular clustering

Fig.1 shows fluctuations in the galaxy counts as measured in the APM Galaxy Survey [5]. Symbols with errorbars show the square root of the variance $\bar{w}_2(\theta)$, e.g. $\Delta N/N \equiv \sqrt{\bar{w}_2}$, measured in squared cells of area $A \equiv \theta^2$. The data can be described by a power law: $\Delta N/N \equiv \sqrt{\bar{w}_2} \simeq (A/A_0)^\beta$ with $A_0 \simeq 7.6 \times 10^{-5}$ sq.deg. and $\beta \simeq -0.175$, for $A < 2.7$ sq. deg. For larger areas there is an exponential cut-off with a characteristic scale of $A_c \sim 110$ sq. deg. The complete model is represented by:

$$\frac{\Delta N}{N} \simeq (A/A_0)^\beta \, \exp[-A/A_c], \tag{1}$$

and is shown as a solid curve in Fig.1.

The effects of shot-noise on the counts, *e.g.* due to the small number of sources in the sample, is easy to quantify. If the mean number density of sources at a given flux is $\mathcal{N}$, then the number of sources in a sample of area A is: $N = A\mathcal{N}$. The total variance on such area is:

$$\bar{w}_2^{total} = \bar{w}_2 + \frac{1}{N}, \tag{2}$$

where $\bar{w}_2$ is the intrinsic variance (in the case of high density) [1]. Thus, the number counts variations due to intrinsic clustering and shot-noise is:

$$\frac{\Delta N}{N} \simeq \left[\bar{w}_2[\,\theta; z; \mathcal{D}] + \frac{1}{A\mathcal{N}} \right]^{1/2} \tag{3}$$

where $\bar{w}_2[\,\theta; z; \mathcal{D}]$ is the angular variance of depth $\mathcal{D}$ at redshift z. We next need to quantify the effects of projection and clustering evolution.

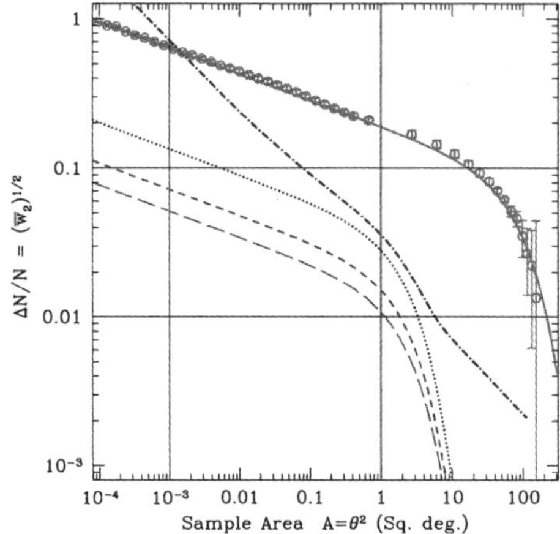

Figure 1. Mean square root deviation in number counts as a function of the sampled area in the sky. Circles with error-bars correspond to counts in square cells of different areas measured in the APM survey. The solid curve shows the model in Eq.[1]. The predicted fluctuations, scaled to a depth of $\mathcal{D} \simeq 2500h^{-1}\mathrm{Mpc}$ ($z \sim 3$), are shown for different clustering evolution models: fixed in co-moving coordinates (dotted line), stable clustering (dashed lines) and linear theory (long-dashed lines). The dot-dashed line includes the shot-noise contribution for a surface density of 2000 sources/sq. deg.

Let us first assume that the 2-pt function $\xi_2(r_{12})$ does not evolve in co-moving coordinates: $\xi_2(r_{12}) = \xi_2(x_{12})$, where $r_{12} = x_{12}/(1+z)$. We then have that:

$$\bar{w}_2(\theta) = \int_V d\boldsymbol{x}_1 d\boldsymbol{x}_2 \; \psi(x_1) \; \psi(x_2) \; \xi_2(\boldsymbol{x}_{12}), \qquad (4)$$

where $d\boldsymbol{x}_i$ is the co-moving volume, $\psi(x)$ is the normalized probability that a galaxy at a coordinate x is included in the catalogue and V is a cone (or pyramid) of radius (or side) θ and infinite depth (with solid angle equal to the sample area A). In the small angle approximation, $e.g.$ when the transverse distances are much smaller than the radial ones, we can relate the angular clustering of samples projected at two different characteristic co-moving depths $\mathcal{D}_2$ and $\mathcal{D}_1$ by:

$$\bar{w}_2[\,\theta\,;\,\mathcal{D}_1\,] = \frac{\mathcal{D}_2}{\mathcal{D}_1}\,\bar{w}_2[\,\theta\mathcal{D}_2/\mathcal{D}_1\,;\,\mathcal{D}_2\,]. \qquad (5)$$

We take $\mathcal{D}_2 \simeq 400h^{-1}$Mpc as the mean APM depth and $\mathcal{D}_1 \simeq 2500h^{-1}$Mpc as the mean co-moving depth of a typical submm sample (the actual number depends on both the mean redshift and the cosmological parameters, as given by the luminosity distance relation). Thus the curve that fits the APM measurements (solid curve) in Fig.1 must be moved left by a factor $\mathcal{D}_2/\mathcal{D}_1$ to account for the fact that, at larger radial distances, the same physical length subtends a smaller angle. The APM curve must also be moved down by a factor $\mathcal{D}_2/\mathcal{D}_1$ because more galaxies are seen in projection at greater radial distances. The resulting prediction in this case, scaled from Eq.[1] to $\mathcal{D}_1 \simeq 2500h^{-1}$Mpc with Eq.[5], is shown as the dotted line in Fig.1. The dotted-dashed line includes the shot-noise contribution in Eq.[3] for a submm 850μm flux limit of 3 mJy, with surface density $\mathcal{N} = 2000$/sq. deg., which are characteristic of the HDF counts. Note how in this case the variance of the counts are dominated by shot-noise at almost all scales. These predictions agree well with the source-count variations found in our submm simulations of the HDF (see Hughes & Gaztañaga in these proceedings), which also have clustering fixed in co-moving coordinates and the same mean depth.

We next model the redshift evolution of ξ_2 in proper coordinates $r_{12} = x_{12}/(1+z)$ as:

$$\xi(r_{12}; z) \simeq (1+z)^{-(3+\epsilon)} \, \xi(r_{12}; 0) \tag{6}$$

For stable clustering (pattern fixed at proper separations) we have $\epsilon \simeq 0$ at small scales [2]. For a power-law correlation with slope $\gamma \simeq 1.7$, linear theory gives $\epsilon = \gamma - 1 \simeq 0.7$ and non-linear growth gives $\epsilon \simeq 1$. If the clustering is fixed in co-moving coordinates, then $\epsilon \simeq \gamma - 3 \simeq -1.3$ which produces even less evolution than in the stable clustering regime. This fixed clustering model describes galaxies which are identified with high density-peaks: peaks move less than particles, which results in less evolution. Weak evolution is consistent with the strong clustering observed in Lyman-break galaxies at $z \simeq 3$, which is comparable to the clustering of present-day galaxies [3]. The above model of cluster evolution projects as:

$$\bar{w}_2 \left[\theta; z_1; \mathcal{D} \right] \simeq \left(\frac{1 + z_2}{1 + z_1} \right)^{3+\epsilon-\gamma} \bar{w}_2 \left[\theta; z_2; \mathcal{D} \right] \tag{7}$$

where γ varies from $\simeq 1.6$ over the smaller scales to $\gamma \simeq 2$ near the exponential break. Here we use $z_2 \simeq 0.15$ for the APM depth and $z_1 \simeq 2$ for the depth in sub-mm samples. The resulting predictions, scaled from Eq.[1] to $\mathcal{D}_1 \simeq 2500h^{-1}$Mpc, are shown as the dotted ($\epsilon = \gamma - 3$), short-dashed line ($\epsilon = 0$) and long-dashed lines ($\epsilon = \gamma - 1$) in Fig.1.

2 Discussion

In summary, our recipe for the angular variance of number count fluctuations $(\frac{\Delta N}{N})^2 \equiv \bar{w}_2$ in a galaxy sample of area $A = \theta^2$, mean co-moving depth $\mathcal{D}$ and mean redshift z, is given by:

$$\bar{w}_2[\,\theta; z; \mathcal{D}] \simeq \frac{400h^{-1}\mathrm{Mpc}}{\mathcal{D}} \left(\frac{1.15}{1+z}\right)^{3+\epsilon-\gamma} (\theta/\theta_0)^{2\beta}\, e^{-(\theta/\theta_c)^2} + \frac{1}{\theta^2\mathcal{N}} \quad (8)$$

where $\beta \simeq -0.175$, $\theta_c \sim 10.5$ deg., $\theta_0 \simeq 8.7 \times 10^{-3}$ deg., $3 + \epsilon - \gamma$ is between 0 and 2, depending on clustering evolution, and $\mathcal{N}$ is the mean number density of sources at the given flux that produces shot-noise fluctuations (*e.g.* Eq.[3]). Several cases for the above model are shown in Fig.1. Assuming the shot-noise is negligible, *e.g.* $\mathcal{N} \to \infty$, to reach a 1% level of fluctuation in N (lower horizontal line in the figure) we require a sample of about $A \simeq 6$ sq. .deg. for the model with co-moving evolution and about $A \simeq 1$ sq. deg. for a model with strong clustering evolution. In the submm survey of the HDF [4], with $A = 0.001$ sq. deg. (the left vertical line in the Fig.1), we expect $\Delta N/N \simeq 0.2$ in the case of weak clustering evolution and $\simeq 0.05$ for the strong evolution case. Thus, in principle, comparing number counts in a few more submm surveys, similar in area to the HDF, would provide a clear discrimination between clustering evolution models.

Nevertheless in small submm surveys the number counts for bright sources will be low, *e.g.* $N = \theta^2\mathcal{N} \simeq 3$ for $S_{850\mu m} \simeq 3\,\mathrm{mJy}$ in the HDF survey [4], and hence the shot-noise correction dominates and masks any evolution in the clustering. This situation is shown as a dot-dashed line in Fig.1 for the submm HDF survey. In this case, we could still subtract the shot-noise contribution using Eq.[3] although this will introduce large uncertainties. Thus, in the future, it is necessary to extend the surveys to larger areas and/or to lower flux densities in order to have a discriminating measurement of clustering. The Gran Telescopio Milimétrico/Large Millimeter Telescope [6] will be the optimal facility to provide this new generation of deep and wide mm surveys.

References

1. Gaztañaga, E., *MNRAS* **268**, 913 (1994).
2. Gaztañaga, E., *ApJ* **454**, 561 (1995).
3. Giavalisco, M. *et al*, 1998 *ApJ* **503**, 543 (1998).
4. Hughes, D.H. *et al*, *Nature* **394**, 241 (1998).
5. Maddox, S. J. *et al*, *MNRAS* **242**, 43P (1990).
6. http://www.lmtgtm.org/

CLUSTER ENVIRONMENTS IN THE EARLY UNIVERSE: PROBING OBSCURED PROTO-ELLIPTICALS WITH SCUBA

ROB IVISON

ATC, Royal Observatory, Blackford Hill, Edinburgh EH9 3HJ
E-mail: rji@roe.ac.uk

IAN SMAIL

Dept of Physics, University of Durham, South Road, Durham DH1 3LE
E-mail: ian.smail@durham.ac.uk

JAMES DUNLOP

IfA, University of Edinburgh, Blackford Hill, Edinburgh EH9 3HJ
E-mail: jsd@roe.ac.uk

CLARE JENNER

Physics & Astronomy, University College London, London WC1E 6BT
E-mail: cej@star.ucl.ac.uk

We describe a survey of high-redshift radio galaxies and quasars, 12 of which have so far been mapped to robust 850-μm detection limits of 5 mJy. Our aim is to test the proposal that these AGN act as signposts to high-density environments in the early Universe, regions that may subsequently evolve into the richest present-day clusters. By searching these fields for dusty starbursts — the progenitors of cluster ellipticals — we can constrain both galaxy- and structure-formation models. Our SCUBA maps reveal an over-density of luminous submm galaxies in the vicinity of high redshift AGN. These sources are often associated with extremely red objects (EROs; $I - K > 5$); the detection of multiple components close to the central AGN, often blended, is also a common theme. These maps thus provide a unique graphical demonstration of the formation of massive galaxies in the early Universe — a process obscured from view to all but SCUBA. We introduce a classification scheme for the counterparts to submm galaxies, based on the results of follow-up observations. Finally, we propose that the over-densities of both submm and ERO sources in these fields represent young dusty, starburst galaxies forming within proto-clusters centred on the massive AGN hosts, regions that are sometimes also traced by a less-obscured population of Lyman-break galaxies.

1 Survey Rationale

1.1 Theory

Biased galaxy-formation theories predict that massive galaxies at high redshifts should act as signposts to high-density environments. These subsequently evolve into the cores of the richest clusters seen at the present day and might be expected to be characterised by over-densities of young galaxies, including a population of dusty, interaction-driven starbursts — the progenitors of massive cluster ellipticals. Confirming that high-redshift radio galaxies reside in high-density environments would also provide substantial insight into galaxy evolution in these regions, especially the formation of giant ellipticals.

1.2 Practicalities

After struggling with the first batch of photometry data from SCUBA[1] — a painful effort to convince ourselves that a coadded flux for 8C 1435+635 that bounced around from one night to the next could in fact be trusted to the required sub-mJy accuracy — we became converts to map64.t, SCUBA's jiggle-mapping mode. Even if it did take umpteen times longer to reach the required flux levels, seeing a beam-sized source appear at the expected position gave tremendous faith in a source's reality and the potential for serendipitous detections was too tempting to pass up, particularly at a time when submm cosmology could boast a measly dozen submm-selected galaxies[2,3,4].

Subsequent efforts to nail high-redshift radio galaxies with SCUBA involved both photometry[5,6] and mapping[7]. In the mapping project, serendipitous detections came thick and fast; the rate of detections far exceeded those of blank-field surveys and it quickly became apparent that there were tentative over-densities of submm sources in the fields of many distant radio galaxies.

For targets at $z \geq 2$ the resolution and field of view of SCUBA at 850 μm means we are sensitive to star-forming galaxies distributed on scales from 100 to 1000 kpc — well-matched to the predicted virial radii of the most massive clusters at these epochs.

2 Initial Results

It seems quite plausible that by mapping high-redshift AGN we have stumbled upon the population of clustered submm galaxies predicted by theoretical work. To quantitatively test this we have undertaken searches in the fields of a dozen radio galaxies and quasars at $z = 3 - 5$, with another 3 fields slated for 2001. Our maps are extremely deep — often down at the mJy rms level at 850 μm — and they reveal a significant over-density of luminous submm galaxies compared to typical fields. For the 4C 41.17 field (not completely representative, but certainly not unique) the likelihood of finding such an over-density in a random field would be <0.002.

If the redshifts are the same as those of the signpost galaxies — an aspect of the study that we are working hard on, via deep imaging with OVRO, IRAM PdBI, VLA, WHT, UKIRT, VLT, Gemini and Keck — then they have bolo-

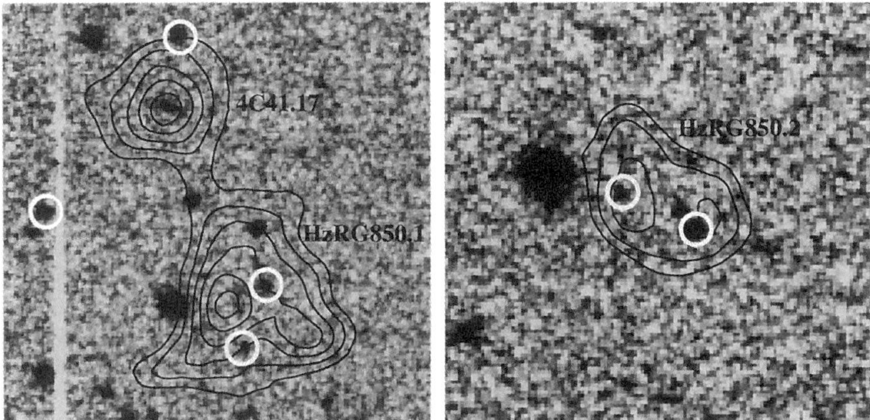

Figure 1: K'-band images of regions around 4C 41.17. 850-μm data are shown as contours. Solid circles denote EROs ($I - K > 5$). The observed 450- to 850-μm flux ratios for 4C 41.17 and HzRG850.1 are 3.2 ± 0.8 and 2.2 ± 0.6, consistent with their being at the same redshift, with best-fit values $z \sim 3.8$ and $z \sim 4.8$. It is interesting that success in obtaining candidate identifications, in particular the association with EROs (Class I sources), correlates with the observed flux density of the submm emission. Is it possible that the brightest submm sources are typically associated with Class I or II counterparts (e.g. HR10, SMM J09429+4658, SMM J02399−0136, SMM J14011+0252, SMM J00266+1708 as well as the brightest sources found by the UK's 8-mJy survey, the Dutch cluster survey and the latest survey with MAMBO) whereas fainter submm sources typically have Class 0 counterparts.

Table 1: Proposed classification of counterparts

Class	Optical/IR Magnitude	Description	Example
0	$I > 26$ and $K > 21$	No plausible counterpart	HDF850.1[3]
I	$I > 26$ and $K \leq 21$	$I - K > 5$ (ERO)	SMM J09429+4658[14]
II a	$I \leq 26$ and $K \leq 21$	Pure starburst	SMM J14011+0252[12]
II b	$I \leq 26$ and $K \leq 21$	Type-II (narrow-line) AGN	SMM J02399−0136[11]
II c	$I \leq 26$ and $K \leq 21$	Type-I (broad-line) AGN	SMM J04135+1027[13]

metric luminosities $L > 10^{13}\,\mathrm{L_\odot}$, which implies star-formation rates consistent with those required to form a massive galaxy in a fraction of a Gyr.

Our target fields also appear to exhibit over-densities of extremely red objects (EROs), some of which may be associated with the submm sources (Figure 1), and Lyman-break galaxies. In a paper detailing the acquisition and analysis of our data in the 4C 41.17 field[7] we have proposed that the over-densities of both submm and ERO sources in these fields represent young dusty, starburst galaxies forming within proto-clusters centred on the high-redshift radio galaxies and quasars, clusters that are also traced by a less-obscured population of Lyman-break galaxies.

3   Nomenclature for counterparts

We have introduced a nomenclature[7,8] for the classification of counterparts to submm galaxies, analogous to that used for proto-stars[9] and building upon the proposed evolutionary scheme for ULIRGs[10]. For operational purposes we based this scheme on the typical depths achieved in typical initial follow-up observations ($I \sim 26$, $K \sim 21$) and the properties of typical submm galaxies[11,12]. The scheme is summarised in Table 1.

We have defined the following classes: Class 0, very-highly obscured sources, where there is no plausible optical or near-IR counterpart; Class I, highly obscured sources, where only a near-IR counterpart exists (often EROs); and Class II, where an obvious optical counterpart is seen (IIa: pure starburst; IIb: type-II AGN; IIc: type-I AGN). The latter class may overlap with the most massive examples of Lyman-break objects. Note that where ultra-deep IR spectroscopy exists it may become possible to identify Class 0/I objects of a particular sub-type (e.g. the recently identified $z = 2.5$ Class Ib galaxy[15,14]).

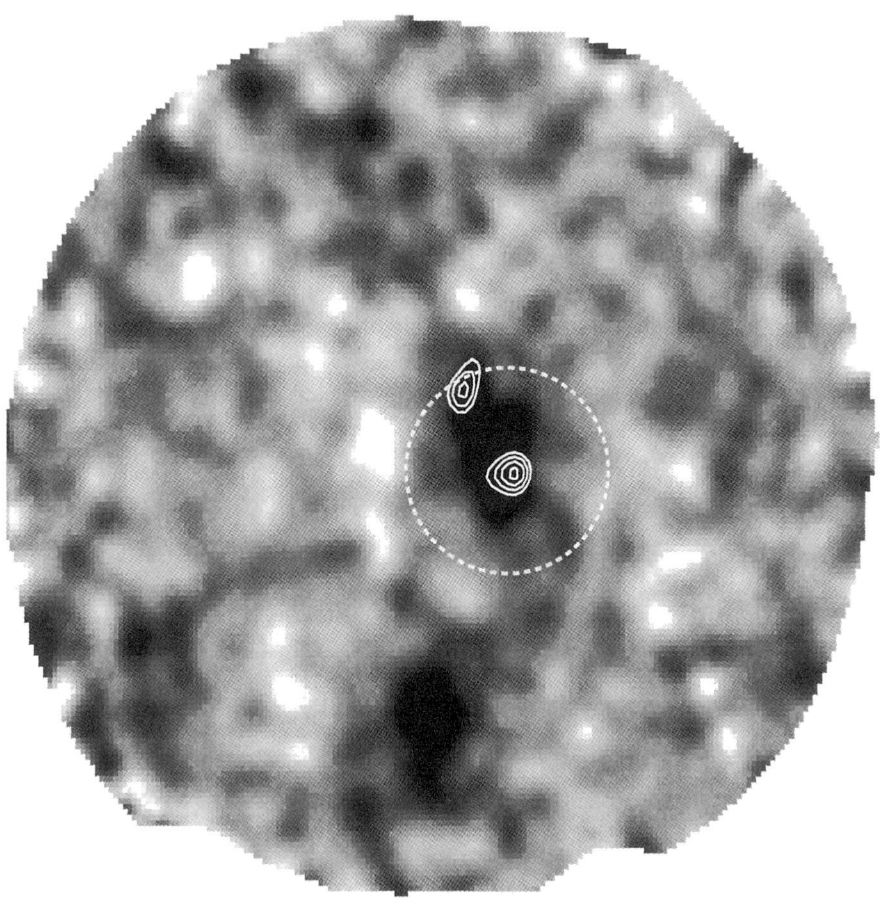

Figure 2: 850-μm image of 6C 1909+72, superimposed with contours from a PdBI CO(4–3) map. The central galaxy is resolved into at least three components by the 14″ SCUBA beam, and another very bright submm galaxy is seen to the south-south-east. Negative features are due to the nodding/chopping procedure (30″ east-west in this case). The CO image is the sum of the channels in the data cube where the emission to the north-east appears strongest, using a naturally-weighted beam and 250 iterations of CLEAN (CO from the radio galaxy itself is thus under-represented). The large circle shows the half-power radius of the PdBI primary beam at 3 mm. No primary beam correction has been applied.

4 The Expanded Sample

Since the first maps of 4C 41.17 were obtained[7], we have expanded the sample to 12 fields, all either radio galaxies or radio-quiet quasars at $z > 3$, each map requiring of order 2–3 shifts of top-grade weather.

Several fields show over-densities of submm galaxies — on average, a factor ~ 3 over the latest blank-field counts. In the early days of the project, putting this over-density on a secure statistical footing was our major goal, together with a desire to obtain redshifts for several of the sources so that CO data could be obtained. In fact, the choice of sample was moderated in part by a desire to map their CO(1–0) emission with the VLA, something we have yet to accomplish, though we are making excellent headway with other samples[18].

On smaller scales ($\sim 100\,$kpc), we were also hopeful of revealing the morphological signature of the intense multiple merging expected during the formation of massive ellipticals in the hierarchical model, providing much-needed information about the presence of merging structures within the halos of the radio galaxies and quasars.

Indeed, in presenting our results for the 4C 41.17 field we noted in passing that several of the bright sources appeared to be resolved, though it was always going to be difficult to prove beyond doubt that the extended nature of the sources was real and that, if real, this was not due to confusion with line-of-sight field galaxies at these faint flux limits.

Putting the over-density of sources aside, the most remarkable feature of the current sample is that 75% of those fields where the signpost AGN is detected show multiple submm components, even at the 14″ resolution of SCUBA. This is not an artefact of poor pointing: in deep, contemporaneous images, the submm-bright quasars BR 1202−0725 and APM 08279+5255 are completely unresolved.

Until now, we have been loathe to comment too much upon the apparently blended appearance of many sources detected with SCUBA. Having seen the effect so often in these maps, however, we are now confident that we can interpret this as clear evidence for intense multiple merging. *The maps thus provide a unique graphical demonstration of the formation of massive galaxies in the early Universe — a process obscured from view to all but SCUBA.*

A particularly good example of this phenomenon is the field of 6C 1909+72 (Figure 2), a radio galaxy at $z = 3.54$ where both dust and CO had already been detected[17] prior to our SCUBA imaging.

Our deep 850-μm image of the 6C 1909+72 field[16] is very-much reminiscent of 4C 41.17. The radio galaxy is detected, together with a very bright source to the south. This time, however, the central galaxy is very clearly resolved, with at least three components detected at the ≥ 5-σ level.

Moreover, one of the 6C 1909+72 submm components is *tentatively* and *serendipitously* detected in CO(4–3) in the PdBI data[17] (which have been re-mapped to highlight the emission to the north-east, close to the half-power point of the PdBI primary beam). We are in the process of confirming the reality of this CO.

If real, the detection of CO at $z = 3.54$ confirms that the component to the north-east is physically associated with the radio galaxy, probably one of many fragments involved in the hierarchical growth of a massive elliptical galaxy. For a CO detection to be possible at the half-power point of the primary beam, this proto-galactic fragment must be extremely massive and gas rich, an interesting development in the struggle to understand the formation and evolution of massive cluster ellipticals.

Acknowledgments

We thank the dozens of friends and collaborators without which this project would have been unthinkable, in particular David Frayer, Thomas Grève, Dennis Downes, Chris Packham, Arjun Dey, Michael Liu, James Graham, Richard Ellis, David Hughes, Huub Röttgering, Wil van Breugel, Michel Reuland, Rich Townsend, Suzie Scott, Chris Carilli, Jean-Paul Kneib and Andrew Blain. Thanks also to the organisers of this timely and exciting conference.

1. Holland W.S. et al., MNRAS **303**, 659 (1999)
2. Smail I., Ivison R.J., Blain A.W., ApJ **490**, L5 (1997)
3. Hughes D.H. et al., Nature **394**, 241 (1998)
4. Barger A.J. et al., Nature **394**, 248 (1998)
5. Archibald E.N. et al., MNRAS, in press (2001)
6. Dunlop J.S., these proceedings, (2001)
7. Ivison R.J., Dunlop J.S. et al., ApJ **542**, 27 (2000)
8. Smail I., Ivison R.J., Blain A.W., Kneib J.-P., in prep, (2001)
9. Andre P. et al., ApJ **406**, 122 (1994)
10. Sanders D.B. et al., ApJ **325**, 74 (1988)
11. Ivison R.J., Smail I. et al., MNRAS **298**, 583 (1998)
12. Ivison R.J., Smail I. et al., MNRAS **315**, 209 (2000)
13. Knudsen K.K., van der Werf P., Jaffe W., these proceedings, (2001)
14. Smail I., Ivison R.J. et al., MNRAS **308**, 1061 (1999)
15. Frayer D.T. et al., in prep, (2001)
16. Ivison R.J., Jenner C.E., Röttgering H.J.A. et al., in prep, (2001)
17. Papadopoulos P., Röttgering H.J.A. et al., ApJ **528**, 626 (2000)
18. Papadopoulos P., Ivison R.J. et al., Nature **409**, 58 (2001)

7. Sub-mm/mm Observations of Known (Low- and) High-Redshift Sources

THE SCUBA-BRIGHT QUASAR SURVEY (SBQS): THE $Z > 4$ SAMPLE

KATE G. ISAAK[1], ROBERT S. PRIDDEY[2], RICHARD G. MCMAHON[2], ALAIN OMONT[3], PIERRE COX[4], STAFFORD WITHINGTON[1]

[1] *Cavendish Laboratory, University of Cambridge;* [2] *Institute of Astronomy, University of Cambridge;* [3] *IAP, Paris;* [4] *IAS, Orsay*

e-mail: isaak@mrao.cam.ac.uk;rpriddey@ast.cam.ac.uk;rgm@ast.cam.ac.uk; omont@iap.fr;cox@ias.fr

We present initial results of a new, systematic search for indicators of massive star formation in the host galaxies of the most luminous and probably most massive $z > 4$ radio-quiet quasars ($M_B \leq -27.5$; $\nu L_\nu(1450A) > 10^{13} L_\odot$). Our $850\mu m$ survey using SCUBA at the JCMT has been designed with a $3\sigma = 10$mJy flux limit in order to identify sources suitable for follow-up studies of high-redshift star formation using molecular spectral line and continuum diagnostics with existing observing facilities. The program has been running serendipitously in the backup queue for 6 months, taking advantage of the short periods when the atmospheric transmission at zenith is between 60% – 75%. We have detected 8 submillimetre-bright sources ($S_{850\mu m} > 10$mJy, from a total of 38 observed) at a significance of 3σ or better. The weighted mean of the flux measured from undetected sources is $\sim 2.0 \pm 0.6$mJy, which we suggest as an underlying fiducial flux for the $z > 4$ radio-quiet quasar host galaxy population.

1 Motivation

Our understanding of star formation at early epochs is due, to a great extent, to the results of studies at submillimetre and millimetre wavelengths. Using (sub)millimetre-wave continuum [9,11,14] *and* the molecular spectral line emission [13,12,3,15] one can start to infer the dust temperature, dust mass, far-infrared luminosity, star formation rate and molecular gas reservoir respectively, and so probe the physical conditions in massive starbursts in the high-redshift Universe. To date, dust and CO emission have been detected from only a handful of $z > 4$ objects – observations, specifically searches for CO, are time-consuming. Thus which of the inferred properties are typical and which are exceptional is not clear. More submillimetre-bright objects need to be identified and studied, but how? In Figure 1 we plot a selection of objects over a range of redshifts for which either dust or/and CO emission have been detected – all objects for which both have been detected have $850\mu m$ fluxes $S_{850\mu m} > 10$mJy. Here, we present the interim results of a survey to find more of such objects, in order to define a statistically significant sample ($N \geq 3\sqrt{N}$) with which to study star formation at high redshift.

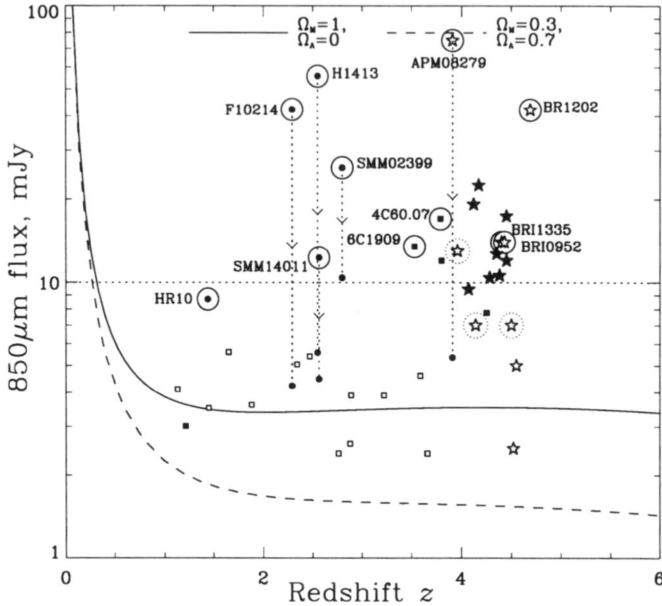

Figure 1. A plot of $850\mu m$ flux vs. redshift for a selection of high-z objects. Stars – detections of $z > 4$ radio-quiet quasars; solid – detections presented in this paper; open – detections from previous work; squares – detections of radio galaxies: unfilled – from Archibald *et al.*; filled from the literature. Rings indicate existing CO observations: solid – detections; dashed – nondetections. Curves show the estimated flux of an Arp220-like object over a range of redshifts (solid line – $\Omega_M = 1; \Omega_\Lambda = 0.$; dashed line $\Omega_M = 0.3; \Omega_\Lambda = 0.7$). The dashed horizontal line denotes the target 3σ limit of this survey

2 The survey

We are using the $850\mu m$ filter on SCUBA [5] at the JCMT to conduct a survey to find the most submillimetre-bright of the brightest optically-selected radio-quiet quasars at $z > 4$. To achieve a sensitivity of $3\sigma \sim 10$mJy at $850\mu m$ requires only $\sim 20 - 30$ minutes of observing time during conditions when the zenith sky transparency is $60\% - 75\%$ (conditions referred to at JCMT as grade-3 weather). Our program, running in contingency-mode, makes use of short periods when the weather is not quite good enough for scheduled programs requiring the highest possible sky transparency – a backup program. Targets are chosen by the "displaced observed" from our source list. Obser-

Table 1. Table of observed and derived properties for detections in the SBQS $z > 4$ survey

Source name	z	M_B	$S_{850\mu m}$ (mJy)	M_{BH} ($10^9 M_\odot$) (1)	$M_{\rm sph}$ ($10^{11}\ M_\odot$) (2)	M_D ($10^9\ M_\odot$) (3)	L_{FIR} $10^{13}\ L_\odot$ (4)	SFR $\Psi \frac{M_\odot}{yr}$ (5)
PSS J0452+0355	4.38	−27.6	10.6	3.6	7.3	0.9	1.2	1200
PSS J0808+5215	4.45	−28.7	17.4	10.0	20	1.5	2.0	1000
PSS J1048+4407	4.38	−27.4	12.0	3.0	6.0	1.1	1.3	1300
PSS J1057+4555	4.12	−28.8	19.2	11.0	22	1.8	2.3	2300
PSS J1248+3110	4.35	−27.6	12.7	3.6	7.3	1.1	1.4	1400
PSS J1418+4449	4.28	−28.6	10.4	9.1	18	0.9	1.2	1200
PSS J1646+5514	4.35	−28.7	9.45	10.0	20	0.8	1.1	1100
PSS J2322+1944	4.17	−28.1	22.5	5.8	12	2.0	2.6	2600

(1) Black hole masses, M_{BH}, calculated assuming an $M_B \rightarrow L_{bol}$ bolometric correction of 12.0 (Elvis *et al.*), and that the AGN is radiating at the Eddington limit; (2) Spheroidal mass, M_{sph}, calculated assuming $M_{sph} \sim 200 M_{BH}$; (3) dust mass, M_D, evaluated assuming $T_D = 40K$, $\beta = 2$ (from Priddey & McMahon submitted 2000), and a dust emissivity of $\kappa_{125\mu m} = 1.875 m^2 kg^{-1}$, extrapolated to an observed wavelength of $850\mu m$ (Hildebrand, 1983); (4) far-infrared luminosity; (5) star formation rate calculated using SFR $= \Psi \times 10^{-10} L_{FIR}$ and $\Psi = 1$.

vations, using the standard photometry mode, typically consist of $30 - 50$ 18s integrations, with the termination criterion defined as $1\sigma \sim 3.3$mJy. Data are reduced using the SURF [6]/ORAC-DR [7] data reduction package. Sky noise is removed using time series of the median-value bolometer. Corrections for sky transmission are made using $850\mu m$.sky-dips. Calibration is achieved using planets and standard secondary calibrators. Observations of the same object are concatenated, and data clipped at 3σ.

3 Results and Discussion

The survey has been running for 6 months. We have detected 7 new submillimetre-bright quasar host galaxies, and have confirmed a previously ambiguous detection of PSS J1048+4407[3] (Table 1). It is interesting to note that while three of these newly detected sources have $S_{850\mu m} > 15$mJy, only one is as bright as BR B1202-0725.[a] There is some suggestion of an upper limit to the observed $850\mu m$ flux, however it is unclear as to what is its origin. Properties derived from both the quasar luminosity and the detected submillimetre flux are tabulated in Table 1, together with a description of how each

---

[a] Here we compare the measured fluxes with 1/2 of that measured for BR B1202-0725, as high-resolution observations at 1.3mm [13] revealed the presence of two distinct, and equally bright sources.

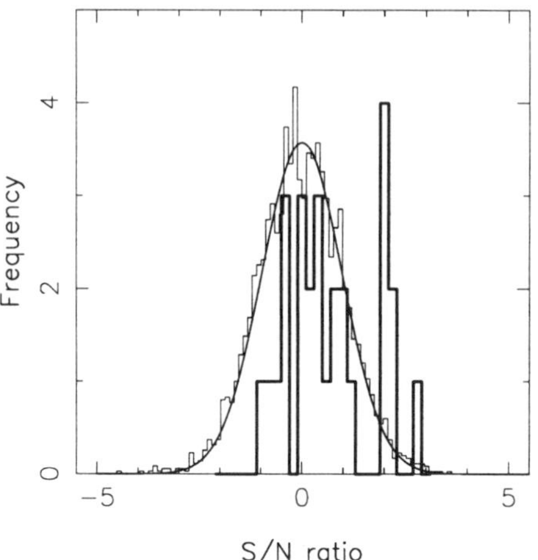

Figure 2. Histogram of the S/N ratios of all $S/N < 3$ sources in our sample: heavy-lined histogram denotes S/N ratios evaluated for the on-source pixel. Superposed (light-line) is a histogram of the S/N distribution for working off-source pixels, over which is plotted a Gaussian approximation to the off-source distribution. The mean of the off-source distribution is zero, indicating that the sky-noise is being removed effectively.

quantity is derived. The dust masses, far-infrared luminosities and inferred star formation rates are all large in comparison with local ultra-luminous galaxies (ULIRGs). Here we assume that the UV emission absorbed by dust and re-radiated in the FIR originates from star formation, rather than from the central AGN. The far-infrared to radio continuum correlation seen in local starburst galaxies in which star formation is the predominant energy source is also observed in BR B1202-0725, as well as a handful of other high-z quasars, by Yun *et al.* Are we, then, observing star formation in objects that are simply scaled up versions of ULIRGs, or are they quite different? Using the locally-derived relationship between the central black hole and spheroidal bulge mass[8] it is clear that we are studying star formation in the early stages of evolution of what turn out to be some of the most massive present-day galaxies – the high-z counterparts of giant ellipticals such as M87.

Non-detections: With a total of 38 observed sources ($3\sigma < 12\text{mJy}$), it is possible to start to consider the redshift and magnitude distribution of both detections and non-detections. Absolute B-band magnitudes have been derived from measured R-band APM magnitudes. At $z \sim 4$ the R-band filter includes the $Ly\alpha$ line, and thus the measured flux is artificially boosted, whilst at higher redshifts the flux is suppressed by absorption due to the $Ly\alpha$ forest. A mean k-correction has been applied to compensate for these effects. Ideally, one would use K magnitudes – an almost direct measure of the rest frame B-band – however, at present 2MASS magnitudes have been released for only a subset of our sample. We leave a more detailed discussion of the magnitude derivation and uncertainties to a future publication. There is suggestive evidence that there is a difference between the cumulative distributions of the detections and the non-detections, however the difference with the current sample is not large enough to reject the null hypothesis that the detections and non-detections are drawn from the same population.

Given the relative homogeneity of the sample, we can use the signal-to-noise (S/N) ratios of the non-detected sources to try to infer a *statistical* measure of the flux of the underlying quasar host galaxy population. For the purposes of this sample we consider all sources that have been observed to $3\sigma < 12\mathrm{mJy}$, thereby including 90% of observed sources. Shown in Figure 2 is a histogram of the S/N ratios of the undetected sources in the sample – clearly bimodal. This is due in part to two of six observations with $S/N > 2$ that were terminated before the nominal $3\sigma \sim 10\mathrm{mJy}$ survey sensitivity was reached – if continued, these two observations would most likely have been detections with $S_{850\mu m} \sim 10\mathrm{mJy}$. The weighted mean of all non-detections with $3\sigma < 12\mathrm{mJy}$, including those with $S/N > 2$, is $2.0 \pm 0.6\mathrm{mJy}$, a value that is consistent with that deduced by McMahon *et al.* [10] using a quite different approach. As a control, we determined the S/N distribution of all "quiet" off-source pixels. As one might expect, the distribution was almost Gaussian with a mean of zero, and so the offset seen with the on-source measurements is a real effect rather than an artifact of the data-reduction process.

4 Future Work

Followup observations of the brightest detections are underway, both continuum observations to constrain the spectral energy distribution and CO line searches to measure the molecular gas mass. Two parallel surveys are currently underway: one based on the program described here, a $z > 4$ survey using MAMBO at IRAM-30m, and the second a survey to define a statistically significant sample at $z \sim 2$. The time interval between these redshifts is ca. 1.5 Gyr, a time scale considerably longer than the e-folding timescale for Eddington-limited accretion onto a massive black hole. Similar detection rates would imply that bulge formation and quasar activity are coeval.

150

Acknowledgements: We acknowledge the displaced observer, Gerald Moriarty-Schieven and the JCMT T.S.S.s. KGI/RSP and RGM acknowledge support from PPARC and the Royal Society respectively. The JCMT is operated by JACH on behalf of PPARC (UK), NRC(Canada) and the SRON(Netherlands).

References

1. Archibald *et al.*, astro-ph/0002083 , (2000)
2. Elvis *et al.* ApJS **95**, 1 (1994)
3. Guilloteau *et al.* A&A **349**, 363 (2000)
4. Hildebrand QJRAS **24**, 267 (1983)
5. Holland *et al.* MNRAS **303**, 659 (1999)
6. Jenness & Lightfoot STARLINK , 216 (2000)
7. Jenness & Economou STARLINK , 231 (2000)
8. Magorrian *et al.* AJ **115**, 2285 (1998)
9. McMahon *et al.* MNRAS **267**, L9 (1994)
10. McMahon *et al.* MNRAS **309**, L1 (1999)
11. Isaak *et al.* MNRAS **269**, L28 (1994)
12. Ohta *et al.* Nature **382**, 426 (1996)
13. Omont *et al.* Nature **382**, 428 (1996)
14. Omont *et al.* A&A **315**, L10 (1996)
15. Papadopoulos *et al.* ApJ **528**, 626 (2000)
16. Yun *et al.* ApJ **528**, 171 (2000)

HIGH RESOLUTION IMAGING OF ULIRGS

N. Z. SCOVILLE

Caltech
Astronomy Dept. 105-24
Pasadena, CA 91125
E-mail: nzs@astro.caltech.edu

Based on the properties of local ULIRGS, studies of high redshift galaxies undergoing starbursts will require better than arcsec resolution. Critical constraints such as the gas mass fraction (which might indicate the youth of the galaxies) and the existence of multiple nuclei (indicative of heirarchical merging) require both kinematic and spatially resolved imaging. In the luminous infrared galaxies, nuclear starbursts and active nuclei are fueled by extraordinarily large masses of gas and dust concentrated at radii of a few hundred pc by viscous accretion and the torques associated with galactic merging. The nearby ULIRGS are probably excellent analogs of galaxies seen at high redshift during the epoch of galaxy formation and growth. We summarize results from a NICMOS survey of 24 ultra-luminous IR galaxies together with mm-interferometry of the molecular gas in these galaxies. Eight of the 24 galaxies imaged with NICMOS have nuclear point sources and eleven have double nuclei. Nine of the 24 systems are fit better by an $r^{1/4}$ law (rather than an exponential disk), suggesting that the young starburst population can relax rapidly in violent mergers. In the IR galaxies much of the enhancement in the star formation probably occurs via the collision of massive clouds since, often, large numbers of bright clusters may be found in the overlap regions of the colliding galaxies. The high resolution NICMOS and mm-wave interferometric imaging of the local ULIRG prototype, Arp 220 indicate double nuclei separated by ~ 350 pc, each of which is embedded in a massive gas and dust disk with typical column densities corrresponding to $A_V \sim 1000$ mag. Studying such systems at high redshift clearly requires mm/submm observations.

1 Introduction

The luminous infrared galaxies are clearly in a phase of *dynamically* triggered evolution – optical/IR imaging of the most powerful galaxies detected in the IRAS survey revealed that virtually all show evidence of a strong interaction (eg. extended tidal tails) or double nuclei [12,6,14,10,17]. And given the large masses of interstellar gas in these galaxies, it is also clear that the progenitors are gas-rich spirals rather than early-type galaxies [16]. Were it not for the replenishment of the ISM in galactic nuclei, both nuclear starbursts and active nuclei (AGNs) would have long ago subsided to lower power levels; resupply of fuel to their central regions of galaxies occurs by highly dissipative, non-circular galactic dynamics such as happens in a galactic interaction. The rich ISM is the ultimate fuel of the prodigious luminosity through either starburst

"

or AGN activity.

In this contribution I summarize some of the recent results from NICMOS imaging of a sample of 24 luminous IR galaxies (10^{11}–4 x 10^{12} $L_\odot$) at 1.1 – 2.2 μm with 0.2″ resolution [22]. I then discuss in detail the prototypical ULIRG system, Arp 220, including mm-wave interferometry results. The lesson from the nearby systems which can probably be carried over to higher redshift objects is the extraordinarily massive ISM and high extinctions concentrated in the nuclei, underscoring the necessitiy of mm/submm imaging.

2 NICMOS Imaging of Local LIRGS and ULIRGS

The sample of 24 galaxies imaged in the NICMOS GTO program [22] range in infrared luminosities from 10^{11} to 4 x 10^{12} $L_\odot$ at λ=8-1000 μm. The majority of the galaxies were taken from the IRAS Bright Galaxy Survey [23]. NICMOS images for 6 of the 24 galaxies are shown as 3-color images in Fig. 1. Five of the galaxies shown are clearly double nuclei systems and three show tidal tails even on the NICMOS images. Bright off-nuclear star clusters, spiral arms, and high reddening in the nuclei may also be seen. All of the 24 galaxies are redder in both $m_{1.1-1.6}$ and $m_{1.6-2.2}$ than unreddened starburst model colors. Their colors clearly require either extincted starlight and/or an AGN energy source but the required extinctions vary tremendously between galaxies and within the individual galaxies [22].

2.1 Luminosity Source and Nuclear Concentration

Considerable *circumstantial* evidence exists relating the luminous infrared galaxies to QSOs. Most significant is the continuity in shape and absolute power of spectral energy distributions between the two classes – specifically, objects may be found with SEDs ranging continuously from those of the most dust-enshrouded ultraluminous infrared galaxies to those of QSOs [14,9,3]. Approximately 50% of the most luminous infrared galaxies exhibit optical emission line ratios indicating a hard-spectrum ionizing source [7,26].

The degree of nuclear concentration of the light as a function of wavelength, luminosity and galaxy type (i.e. optical spectral class or IR warm vs cold colors) can provide important clues to the luminosity source and evolutionary state of the galaxies. All of the eight galaxies with significant nuclear point-source contributions (NGC 2623, NGC 7469, IRAS 08572+39, IRAS 05189-25, PKS1345+12, IRAS 07598+65, Mrk 1014, and 3C48) in the NIC-MOS images are also classified as *warm* in terms of their mid-infrared colors. Similar conclusions can be drawn with respect to the optical spectral classi-

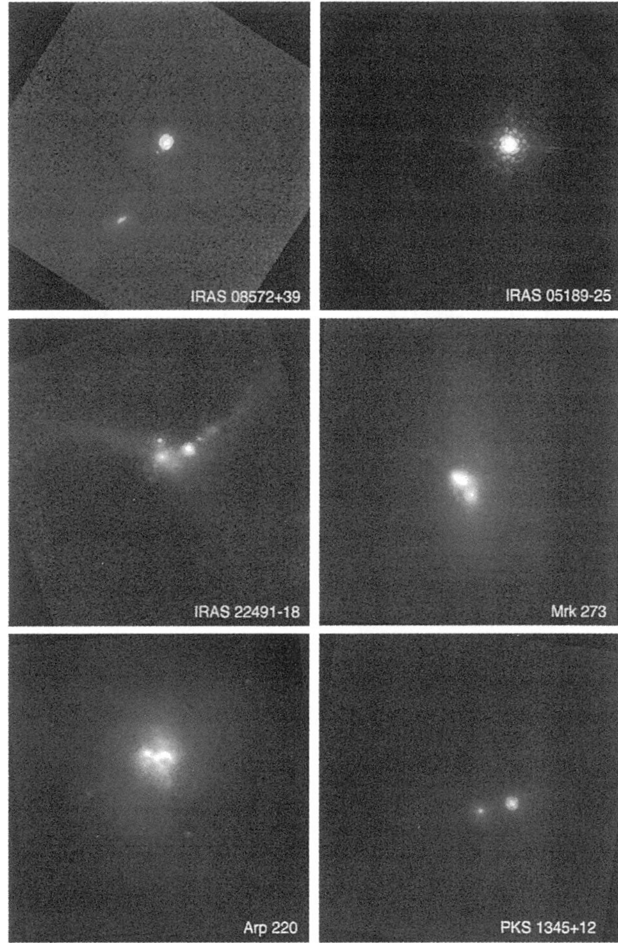

Figure 1. Three color NICMOS images are shown for 6 ULIRG galaxies with far infrared luminosities $1.2\text{--}1.6\times10^{12}$ $L_\odot$. The blue (1.1μm), green (1.6 μm) and red (2.2 μm) were individually log-stretched to bring out the maximum structure and to enhance the observed color gradients. The FWHM are $0.11\text{--}0.22''$ at $1.1\text{--}2.22$ μm respectively.

fications – i.e. most of the galaxies with nuclear point sources contributing significantly in the near-infrared are classified as Seyfert or QSO, yet not all of the galaxies with AGN-like spectra have significant nuclear point sources.

Comparison of the light profiles in the three bands for the 24 NICMOS GTO galaxies clearly shows that in virtually all cases the 2.2 μm flux is more centrally peaked at the nucleus than that at 1.1 and 1.6 μm. In 9 of 24 galaxies (NGC 4418, Zw049.057, NGC 2623, IC883, NGC 6240, UGC 5101, IRAS 10565+2448, Arp 220, and IRAS 14348-1447), the light profiles at 1.6 μm are fit better by an $r^{1/4}$-law than by an exponential disk profile. A smaller percentage is fit by an exponential disk profile. This suggests that the stellar population whose light dominates the inner $5-10$ kpc in these galaxies appears to be better approximated by a spheroidal rather than a disk-like configuration. If the near–infrared light is dominated by young stars such as red supergiants, these stars must have formed during the merger and have already assumed elliptical-like orbits. Whether these systems end up forming giant ellipticals will depend on a number of factors – most importantly, the overall mass density of stars in the central regions and the quantity of ISM left over in a cold disk after the merging is complete. Kormendy & Sanders (1992) point out that in some of the ultra-luminous sytems (eg. Arp 220), the central mass density is in fact similar to that of elliptical galaxy cores if the massive ISM component is included – the presumption is then that an elliptical galaxy could be the end product if a significant fraction of the ISM is converted into stars.

3   Arp 220 – A 'prototypical' ULIRG

Mm-wave imaging provides a unique capability to probe the ISM distribution and kinematics. More than 20 luminous ($\geq 10^{11} L_\odot$) infrared galaxies have now been imaged, primarily at OVRO and IRAM [19,20,4,2,24]. Virtually all display massive concentrations of molecular gas in the central few kpc.

Arp 220, at 77 Mpc, is one of the nearest and the best known ultra-luminous merging system ($L_{8-1000\mu m} = 1.5 \times 10^{12}$ L$_\odot$). Visual wavelength images reveal two faint tidal tails, indicating a recent tidal interaction [6], and high resolution ground-based radio and near-infrared imaging show a double nucleus [1,5]. The radio nuclei are separated by 0.″98 at P.A. $\sim$90° [1], corresponding to 350 pc. To power the energy output seen in the infrared by young stars requires a star formation rate of $\sim$$10^2$ M$_\odot$ yr^{-1}. Arp 220 has been the subject of a number of OVRO and IRAM interferometer studies imaging in the 2.6 mm CO line [19], 3 mm HCN [11], and 1.3 mm CO [20,4,13]. The CO (2–1) line emission, mapped at 1″ resolution, showed two peaks separated by 0.9″,

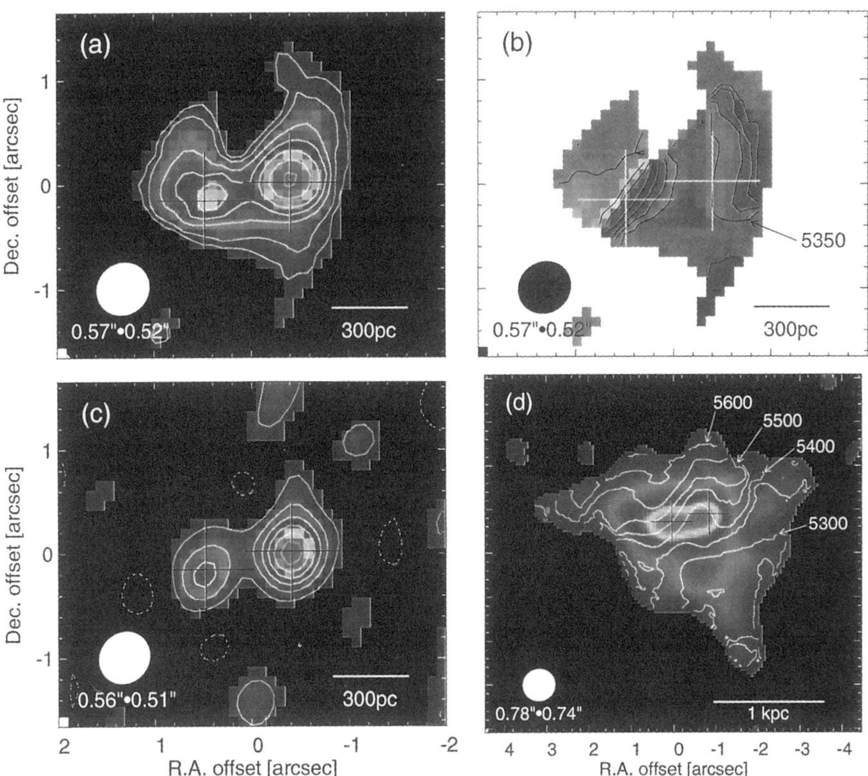

Figure 2. The merging nuclei of Arp 220 are shown in 0.5″ resolution imaging of the CO(2–1) and dust continuum emission. These data clearly resolve the two nuclei and reveal for the first time counter-rotating disks in each nucleus. The panels show : a) continuum-subtracted CO(2–1) (using only high resolution data), b) the CO mean velocities, c) the 1.3 mm dust continnum, and d) the total CO emission including both low and high resolution interferometry (Sakamoto etal 1999). Crosses indicate the 1.3 mm continuum positions of the nuclei.

and an inclined disk of molecular gas [20,4]. These peaks correspond well with the double nuclei seen in near-infrared and radio continuum images. The 0.5″ resolution CO and 1.3 mm continuum maps obtained recently by Sakamoto *et al.*(1999) using OVRO are displayed in Figure 2. These reveal **counter-rotating** disks of gas in each of the nuclei. The kinematic data clearly require very high mass concentrations in each nucleus, consistent with their being individual galactic nuclei. The fact that they are counter-rotating is consistent with the concept that more complete merging may be associated with counter-rotating precursor galaxies in which there can be greater angular momentum cancellation. The masses in each nucleus are apparently dominated by the molecular gas – a common finding of the the ULIRG galaxy studies [2].

The NICMOS images for the central region of Arp 220 clearly show the two nuclear regions plus several lesser peaks and the morphology changes remarkably with wavelength [21]. In particular, the bright western nucleus shows greater extension to the south at 2.2μm, and the eastern nucleus has a southern component which becomes increasingly apparent at the longer wavelengths. Our registration of the near-infrared relative to the radio places one radio nucleus between the two emission peaks seen at 2.2μm in the east and the western radio nucleus lies in the area of extremely high obscuration to the south of the western 2.2μm peak [21]. The crescent or partial ring morphology of the western nucleus might readily arise if there is an obscuring disk of dust and gas embedded in a spheroidal nuclear star cluster.

The total molecular gas content for Arp 220 is $9 \times 10^9 M_\odot$ based on the CO (2–1) emission and a CO-to-H_2 conversion ratio that is 0.45 times the Galactic value [20]. This enormous mass (approximately two times that of the total Galactic ISM) is contained entirely within R <1.5 kpc and approximately $5 \times 10^9 M_\odot$ is apparently concentrated in a thin disk in the nuclear region at radii < 250 pc. The gas probably collected in the center of the merging system as a result of torques associated with the encounter and the high dissipation in the gas. The inferred mean extinction through the disk in the center of Arp 220 is $A_V = 1000$–2000 mag [20].

4   Induced Starburst Activity

In normal spiral galaxies like the Milky Way, most of the star formation proceeds at a steady, low efficiency rate. This may be seen from the fact that estimates for the current star formation rate in the Galaxy are typically $\sim$ few $M_\odot$ yr^{-1} while the supply of star forming molecular gas is $\sim 3\times10^9$ $M_\odot$, implying a gas–star cycling time of 10^9 yrs. Yet, the majority of the molecular gas is at average density 250 H_2 cm^{-3} with a free-fall collapse time of 10^7 yrs. Therefore, on the timescale of gravitational free-fall, the formation of stars occurs with an efficiency of only 1%. It is generally believed that the clouds are supported against collapse by a combination of turbulent motions and magnetic fields. The magnitude of the turbulent or magnetic support can be judged by the fact that the molecular emission linewidths of individual

Galactic GMCs are typically 10 times the thermal speed of sound in the gas (i.e. $M = 10$). We conclude that in order *to change the internal state of a GMC and elevate the rate of collapse to form stars, the disturbance must be highly supersonic and in an external medium of density comparable to that of the molecular gas (unless the Mach number $M >> 10$).*

What better way to *'arouse or wake up'* a supersonically supported cloud than to collide it supersonically with another cloud – the ram pressure resulting from the collision will automatically exceed the internal magneto-turbulent pressure! In principle, the cloud-cloud collision rate will depend quadratically on the local volume density of the clouds, their cross sections, and on the cloud-cloud velocity dispersion. In the spiral arms of a galaxy like the Milky Way, the number-density of clouds is enhanced due to orbit crowding in the arms and the OB star formation rate is possibly enhanced due to cloud-cloud collisions [18]. In interacting and merging galaxies, the cloud collision rate will be enhanced : first in the individual disks due to tidally-elevated velocity dispersion (specifically, the normally circular, 'non-intersecting' orbits are disrupted); later due to the passage of the disks through each other; and lastly due to the concentration of the gas at smaller radii (i.e. smaller volume) in the merger nucleus. The first two effects are clearly evidenced in the NICMOS ULIRG sample – for example many luminous star forming regions can be seen in the tidal tails (eg. IRAS 22491-18) and in the galaxy overlap region (eg. NGC 6090 and VV114), where massive concentrations of molecular gas are also seen [2,28]. It is interesting to note that evidence of greatly enhanced star formation due to molecular cloud collisions is also found in the NGC 4038/39 system (the 'Antennae') since the galaxy overlap region exhibits multiple CO emission velocities along the line of sight precisely coinciding with the location of very strong 15μm emission seen with ISO [27]. An alternative mechanism for the starbursts has recently been suggested [25] in which the pre-existing clouds are removed from their source of internal ionization by cosmic rays, the magnetic fields can then more rapidly diffuse out of the dense gas, and then without the magnetic support the cloud undergo rapid gravitational collapse. This scenario might account for high rates of star formation in merging galactic disks by virtue of the fact that the clouds are temporarily displaced from the galactic disk cosmic rays and magnetic fields.

5 Concluding Remarks

Interactions and merging play a fundamental role in the evolution of galaxies – producing the most luminous starburst galaxies and very likely luminous AGN. The dynamical effects of the interactions are most dramatic in the ISM since it is dissipative and has high filling factors. The torques and increased velocity dispersions due to the galactic encounters lead to rapid transport of the dense molecular ISM to the nuclear regions and high rates of cloud-cloud collisions and/or reduced cloud magnetic support [25]. The shock-induced cloud compression might trigger formation of super starburst clusters in the regions of physical overlap of the galactic disks and in tidal bridges/tails. In many of the ultra-luminous IR galaxies, over 10^9 $M_\odot$ of H_2 gas is found at radii $<<$ 500 pc (comparable with the Galactic ISM mass, but within an area 50 times smaller). Within the central region, it now appears this gas dissipatively settles into a thin disk (on the basis of both high resolution mm-interferometry and NICMOS imaging). These ultra-massive nuclear gas disks are presumably the sites of the nuclear starburst activity – but some fraction of the gas may also accrete inwards to much smaller radii, feeding and building up a central AGN.

These are exciting times for both observations and theory, in as much as one is now given a single root cause to possibly understand the origin of both starbursts and AGN. Galactic merging is of course likely to be more frequent in the early universe when the space density of galaxies was much higher and galaxies had an even larger fraction of ISM mass leading to the dramatic rise in QSO luminosity function and the sub-mm source population at higher redshift. There is the very exciting possibility that the major processes responsible for the early galactic evolution can, in fact, be studied in detail in the nearby luminous infrared galaxies.

Acknowledgments

It is a pleasure to acknowledge the many collaborators in much of this work, including mostly recently K. Sakamoto, M. Yun, D. Frayer, A. Evans, R. Thompson and M. Rieke. The OVRO Millimeter Array is a radio telescope facility operated by the California Institute of Technology and is supported by NSF grant AST 9981546.

References

1. Baan, W. A., & Haschick, A. D. 1995, *ApJ*, 454, 745
2. Bryant, P. M. & Scoville, N. Z. 1999, *AJ*, 117, 2632
3. de Grijp, M. H. K., Miley, G. K., Lub, J. & deJong, T. 1985, *Nature*, 314, 240
4. Downes, D., & Solomon, P.M.. 1998, *ApJ*, 507, 615
5. Graham, J. R. *et al.* 1990, *ApJ*, 354, L5

6. Joseph, R.D. & Wright, G.S. 1985, *MNRAS*, 214, 87.

7. Kim, D.-C., Sanders, D. B., Veilleux, S., Mazzarella, J. M & Soifer, B. T. 1995 , *ApJ Suppl*, 98, 129

8. Kormendy, J. & Sanders, D. B. 1992, *ApJ*, 390, L53

9. Low, F. J., Cutri, R. M., Huchra, J. P. & Kleinmann, S. G. 1988, *ApJ*, 327, L41

10. Murphy, T. W., Armus, L., Matthews, K., Soifer, B. T., Mazzarella, J. M. & Neugebauer, G. 1996, *AJ*, 111, 1025

11. Radford, S.J. *et al.*1991, Proc. of IAU Sym. 146, 'Dynamics of Galaxies and Their Molecular Cloud Distributions', eds F. Combes & F. Casoli, (Dordrecht: Kluwer), 303

12. Rieke, G. H. & Low, F. J. 1972 , *Ap J Letters*, 176, L95

13. Sakamoto, K., Scoville, N.Z., Yun, M.S., Crosas, M., Genzel, R., et al. 1999, *ApJ*, 514, 68

14. Sanders, D.B., Soifer, B.T., Elias, J.H., Madore, B.F., Matthews, K., Neugebauer, G., & Scoville, N.Z. 1988, *ApJ*, 325, 74

15. Sanders, D. B., Scoville, N. Z. & Soifer, B. T. 1988, *ApJ*, 335, L1

16. Sanders, D. B., Scoville, N. Z. & Soifer, B. T. 1991, *ApJ*, 70, 158

17. Sanders, D. B. & Mirabel, I. F. 1999, *Ann. Rev. Ast. & Astro.*, 34 , 749

18. Scoville, N. Z., Sanders, D. B. & Clemens, D. P. 1986 *Ap J Letters*, 310, 77

19. Scoville, N. Z., Sargent, A.. I., & Sanders, D. B. 1991, *Ap J Letters*, 366, L5

20. Scoville, N. Z., Yun, M. S., & Bryant, P. M. 1997, *ApJ*, 484, 702

21. Scoville, N. Z., Evans, A. S., Dinshaw, N., Thompson, R., Rieke, M., Schneider, G., Low, F., Hines, D., Stobie, B., Becklin, E., & Epps, H. 1998, *ApJ*, 492, L107

22. Scoville, N. Z., Evans, A. S., Thompson, R., Rieke, Hines, D., Low, F., Dinshaw, N., Surace, J., & Armus, L. 2000, *AJ*, 119, 991.

23. Soifer, B. T., Boehmer, L., Neugebauer, G. & Sanders, D. B. 1989, *AJ*, 98, 766

24. Tacconi, L. J. 1999 in proc. conf on '3-d Imaging in the Universe' (in press), ed. W. van Breugel.

25. Tohline, J., Scoville, N. Z. and Strong, A. 2000' *ApJ̃*(submitted)

26. Veilleux, S., Kim, D.-C., Sanders, D. B., Mazzarella, J. M., & Soifer, B. T. 1995, *ApJ Suppl*, 98, 171

27. Wilson, C. D., Scoville, N. Z., Madden, S.D. & Charmandaris, V., 1999, *ApJ̃*(submitted)

28. Yun, M. S., Scoville, N. Z. & Knop, R. A. 1994, *Ap J Letters*, 430, 109

KNOWN AND UNKNOWN SCUBA SOURCES

DOUGLAS SCOTT

COLIN BORYS, MARK HALPERN, ANNA SAJINA

Department of Physics & Astronomy, University of British Columbia, Vancouver,
BC V6T 1Z1 CANADA
E-mail: dscott@astro.ubc.ca

SCOTT CHAPMAN

Observatories of the Carnegie Institution of Washington, Pasadena, CA 91101
USA

GREG FAHLMAN

Canada-France-Hawaii Telescope, Kamuela, Hawaii 96743 USA

Summary and discussion of some projects to use SCUBA to target sources selected
at other wavebands, as well as to find new sub-mm galaxies in 'blank fields'.

1 Introduction

The sub-mm waveband has opened up for cosmology, and through this window
we can hope to glimpse some answers to a number of related puzzles: What
are the brightest sub-mm galaxies? What sorts of galaxies make up the Far-IR
Background (FIB)? When did the Universe form the bulk of its stars? How
important is dust obscuration for obtaining a full star-formation census?

SCUBA[1] has been instrumental in establishing this new field of sub-mm
high redshift astronomy. This meeting has been dominated by SCUBA-based
surveys, and stands as a testament to the instrument builders who produced
the best camera of its kind at just the right time.

Many basic questions have already been answered by the SCUBA data
that are in hand, and a coherent picture has developed – typical SCUBA
sources are the $z \sim 2$–3 counterparts of locally well-studied Ultraluminous Infrared Galaxies, and these are considerably more common at an epoch that
was perhaps important for the formation of elliptical galaxies. However, there
are still a great many details to understand about the exact composition of the
SCUBA-bright sources, including how they overlap with populations selected
at other wavelengths, how the $\sim$ mJy sources might differ from the brighter
ones, what the clustering strength of the SCUBA sources might be, and what
fraction could be at still higher redshifts. Here we will avoid reviewing all of
the different studies (many of which are covered in other presentations), but

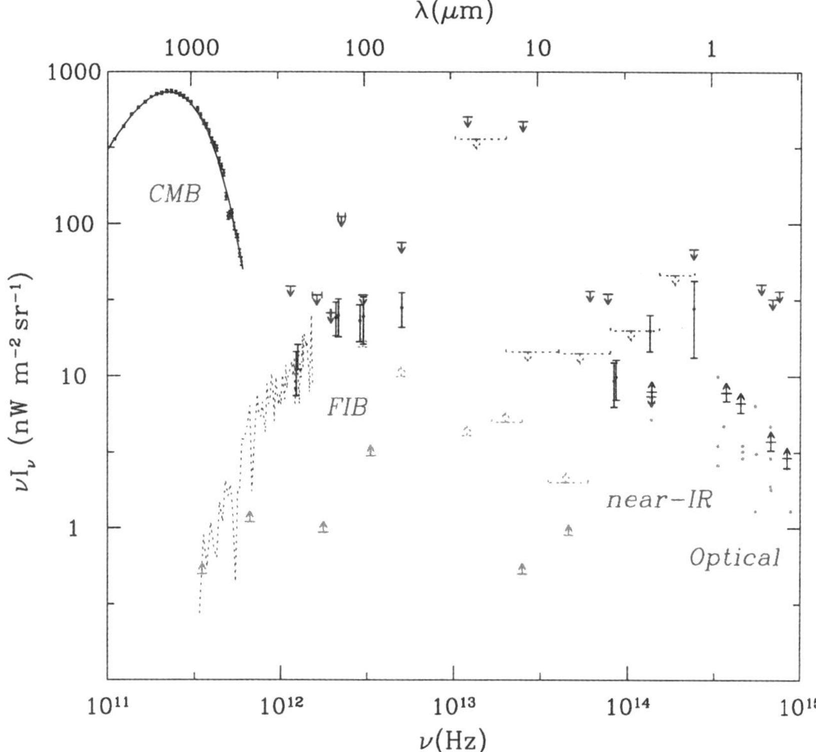

Figure 1. Estimates for the cosmic background in the IR and neighbouring wavebands. This is a blown-up and up-dated plot of Figure 26.2 in *Astrophysical Quantities*. Lower limits generally come from the integrated brightness of resolved sources. Limits that are more model-dependent are indicated by dotted lines. The FIB has been slowly emerging from careful analyses of (largely) *COBE* data. It appears to contain more energy than the optical background (usually assumed to be the sum of the deepest HST counts). It also appears to be broader than the blackbody CMB peak, possibly indicating that it comes from a range of redshifts.

will concentrate on projects carried out by our own group.

We have been undertaking a set of related studies, focussing on either blank fields to detect sub-mm sources, or on characterizing the sub-mm properties of objects selected at other wavelengths.

2 Known sources

2.1 FIRBACK

What are the galaxies that make up the FIB? This question is the main motivation behind the FIRBACK project, which made deep $170\,\mu$m images of the sky with ISO, to break up the background into the sources that produce it[3,4,5,6]. Data from the ISOPHOT instrument are the only currently available information for studying the composition of the FIB *that comes directly from wavelengths where the background peaks*[2]. We have been studying the SCUBA properties of some of these already known FIRBACK-selected sources. Because of the large ISOPHOT beam, this is really feasible only for sources with reliable radio identifications. Photometry with SCUBA allows you to go deeper than mapping in the same integration time. We have also successfully been using '3-bolometer chopping', by choosing the chop position to be approximately coincident with two bolometers in the inner ring of the array, so that those negative signals can be added in, to improve the signal-to-noise[7].

The combination of ISOPHOT + VLA + SCUBA selects galaxies covering a range of redshifts but peaking at $z \sim 1$ (supported by the handful of objects for which redshifts have already been obtained). The study of FIRBACK galaxies at $z \sim 1$ thus bridges the gap between the well-studied $z \sim 0$ ULIRGs and the $z \sim 2$–3 SCUBA sources. At these somewhat more modest redshifts the SEDs of the sources can be realistically constructed over a wide range of wavelengths.

The total number of sources detected in the ELAIS 'N1' field[8] is ~ 120 with $S_{170} > 120\,$mJy. They produce about 10% of the cosmic Far-IR Background at this wavelength, and an entirely unknown fraction of the longer wavelength sub-mm background. Note that SCUBA 'blank fields' tell you only about the background at $850\,\mu$m, where $\nu I_\nu \sim 30$ times lower than its peak value (see Fig. 1). Our first set of results from a sample of 10 objects in the N1 field was presented in Scott et al. (2000)[9]. There we firmly detected 4 sources at $850\,\mu$m. Statistically the sample was detected at 7.5σ at $850\,\mu$m and 4σ even at $450\,\mu$m. Crude photometric redshifts can be obtained simply by comparing the $170\,\mu$m and $850\,\mu$m fluxes, as is shown in Fig. 2.

Joint fits to $L_{\rm FIR}$ and z show that 3 of the galaxies are consistent with luminosities typical of Arp 220, but at $z \sim 1$. The others seem more likely to be at redshifts intermediate between 0 and 1, and intrinsically fainter (but still representing a population that hardly exists locally). Multi-wavelength studies are underway (Lagache et al. in preparation), and the SCUBA work should continue, with a sample covering a wider range of ISO fluxes and

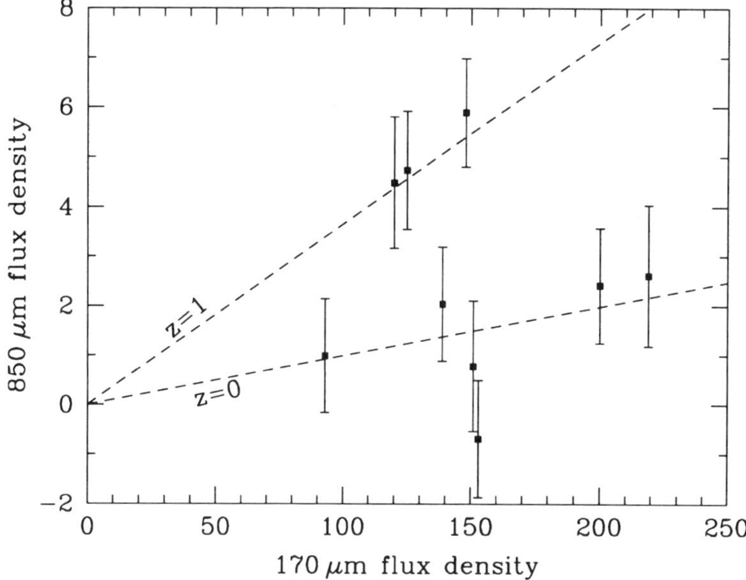

Figure 2. Measured $170\,\mu$m vs $850\,\mu$m flux densities for our first sample of FIRBACK sources. The two dashed lines are $T_{\mathrm{d}} = 50\,$K, $\beta = 1.5$ modified blackbodies at $z = 0$ and $z = 1$. This shows that just the combination of $850\,\mu$m and $170\,\mu$m can give crude redshift information about these galaxies (three seem to lie at higher z than the others) – more detailed SED fitting of the full radio/submm/far-IR/mid-IR/near-IR/optical data will reveal much more of course.

properties at other wavelengths.

2.2 The Blob

Extensive spectroscopic surveys have shown strong clustering among the star-forming Lyman Break Galaxy population at $z \simeq 3$ (which is discussed by Chapman et al. in these proceedings). A high contrast overdensity of LBGs at $z = 3.09$ in the SSA22 region was discovered by using deep narrow-band Ly α imaging to identify at least 160 members of the structure. This region appears to be about 6 times more overdense in LBGs than the general field, and the region centred on the most extended Ly α emission is overdense by roughly a further factor of 2. This is where we centred our SCUBA map[10]. The brightest source in that map (Fig. 3) is $\simeq 20\,$mJy, and centred precisely on the extended Ly α region referred to as 'Blob 1' in Steidel et al. (2000)[11].

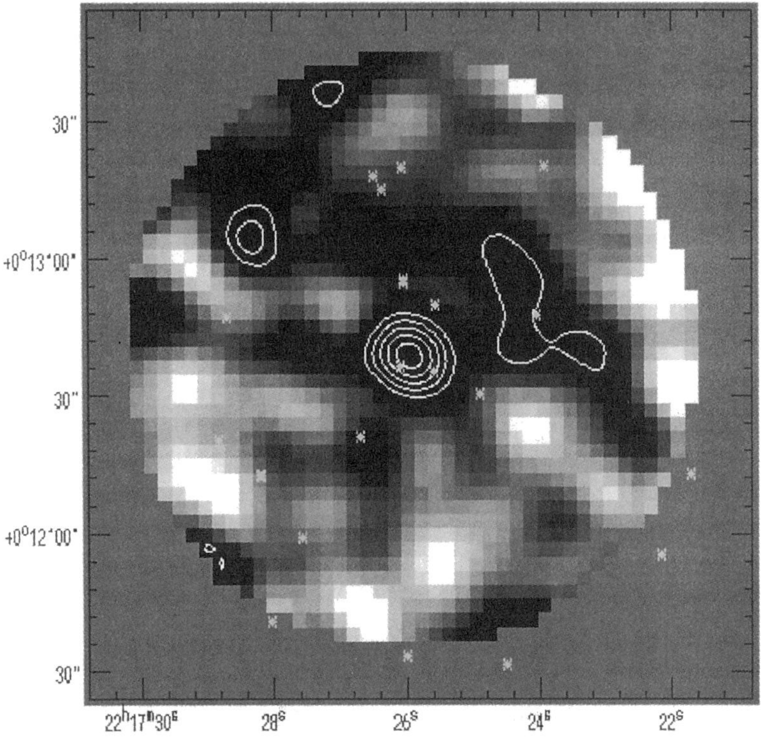

Figure 3. SCUBA 850 μm map centred on 'Blob 1' of the Ly α survey of Steidel et al. (2000). Crosses mark the positions of known $z \simeq 3$ galaxies. Black is positive emission here, and the contours are at 3, 4, 5, 6 and 7σ.

Such sub-mm emission is unexpected, since dust usually implies no Ly α. The simplest explanation is an obscured AGN, which provides the excitation for the extended Ly α, and whose nucleus is responsible for the sub-mm emission. It is also possible that some of the signal could be from the Sunyaev-Zel'dovich increment in a very overdense group, say, forming at $z \sim 3$, but conditions would have to be rather extreme for this to be the case. Nevertheless it is worth following up in detail at other wavelengths to check. This surprising result suggests that the combination of SCUBA and narrow-band Ly α studies could prove to be a fruitful one.

There are some additional sources in our SCUBA map, indicating an overdensity in sub-mm sources, matching that in the LBG population. It would be interesting to map this region more extensively to determine if it is a region of enhanced clustering of SCUBA sources.

3 Unknown sources

We have made reasonably large maps of the Hubble Deep Field Flanking Fields and part of the Groth Strip (see the contribution by Borys et al.) in order to find the brightest SCUBA sources in those well-studied regions. In addition we have a large amount of 'blank sky' data from the off-centre bolometers obtained during photometry observations of known sources. Although too under-sampled to be much use for mapping, these data can nevertheless be studied statistically[12].

3.1 Cluster survey

We have also looked at several rich cluster fields. Here the lensing amplification boosts the number counts (as pioneered by Blain, Ivison, Smail and collaborators). We have collected data on 9 separate clusters: Cl 0016+16, MS 0451-03, Abell 520, Zwicky 3146, MS 1054-03, MS 1455+22, Abell 2163, Abell 2219 and Abell 2261. The clusters were mapped to a variety of depths under different observing conditions. But the resulting maps typically contain a couple of convincing sources. Details are presented in Chapman et al. (2000)[13].

Instead of running through each cluster field in detail, let us focus on only one, Abell 520. Fig. 4 shows our map (upper left), together with an almost overlapping map taken from the SCUBA archive. Each map appears to show 2 detections. There are no obvious near-IR or radio identifications for these sources, but the follow-up data are, as yet, not very deep. One of the sources we discovered (SMM J 04543+0257) appears to be $\simeq 30$mJy, placing it among the brightest 'blank sky' sources.

4 Conclusions

Further clues to the nature of sub-mm sources and how they overlap with optical-, IR- and radio-selected galaxies will be found through the joint approach of investigating the sub-mm properties of already known objects, and finding identifications for possibly unknown objects that are bright in the sub-mm. Studies of the clustering of SCUBA sources are just in their infancy, but should provide further cosmological information. The crucial thing is to iden-

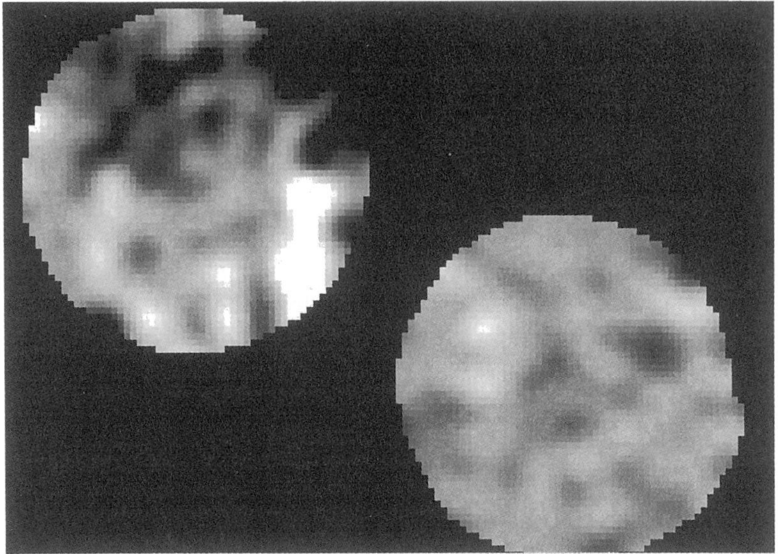

Figure 4. SCUBA 850 μm maps of Abell 520, at two different positions which almost overlap. Here positive emission is white and the noisy edge regions have been clipped. The map at upper left is from our own data and shows two very bright sources at ~ 15 mJy and ~ 30 mJy. The lower right map is from the SCUBA archive, is less noisy, and shows two ~ 10 mJy sources.

tify the SCUBA sources at other wavelengths – this has proved to be difficult in practice, and the sources for which it is *most* difficult are potentially the most interesting!

Acknowledgments

DS wishes thanks all his collaborators on these observational projects, which keep him away from his day job.

References

1. Holland, W.S., et al., MNRAS **303**, 659 (1999).
2. Cox, A.N., 2000, Allen's Astrophysical Quantities, Fourth Edition, Springer-Verlag, New York.
3. Puget J.-L., et al., A&A **345**, 29 (1999).

4. Dole, H., et al., 1999, A&A, in press, astro-ph/9812039.
5. Dole, H., et al., 2000, in 'ISO Surveys of a Dusty Universe', eds. D. Lemke, M. Stickel, K. Wilke, Springer Lecture Notes of Physics, in press, astro-ph/0002283.
6. Lagache, G., et al., 2000, in 'ISO Surveys of a Dusty Universe', eds. D. Lemke, M. Stickel, K. Wilke, Springer Lecture Notes of Physics, in press, astro-ph/0002284.
7. Chapman, S.C., Scott, D., Steidel, C.C., et al., 2000, MNRAS, in press, astro-ph/9909092.
8. Oliver, S., et al., 2000, MNRAS, submitted, astro-ph/0003263.
9. Scott, D., et al., A&A **357**, L5 (2000)
10. Chapman, S.C., et al., 2000, ApJL, in press, astro-ph/0010101.
11. Steidel, C.C., et al., ApJ **532**, 65 (2000).
12. Borys, C., Chapman, S.C., Scott, D. MNRAS **308**, 527 (1999).
13. Chapman, S.C., Scott, D., Borys, C., Fahlman G.G., 2000, MNRAS, submitted, astro-ph/0009067.

A Submillimeter Selected Quasar in the Field of Abell 478

K.K. Knudsen, P.P. van der Werf and W. Jaffe
Leiden Observatory, P.O. Box 9513, NL–2300 RA Leiden, The Netherlands
E-mail: kraiberg@strw.leidenuniv.nl

We report the discovery of a $z = 2.83$ quasar in the field of the cooling flow galaxy cluster Abell 478. This quasar was first detected in a submm survey of star forming galaxies at high redshifts, as the brightest source. We discuss the optical spectrum and far–IR spectral energy distribution (SED) of this object.

1 Introduction

We are performing a survey of gravitationally lensed submm sources — a project aimed at studying the cosmic star formation history. Our survey covers 11 clusters of galaxies and has so far resulted in the detection of at least 25 submm sources. Here we present results on the brightest submm source detected.

2 Observations and Results

In the field of Abell 478 we have detected with SCUBA (JCMT, Hawaii) a very bright submm source, $F_{850\mu m} = 25 \pm 3$ mJy and $F_{450\mu m} = 63 \pm 20$ mJy, and possibly a fainter source of $F_{850\mu m} = 9 \pm 2$ mJy. The errors contain the detection uncertainty and the calibration error. To identify the sources we have obtained a deep I–band image with FORS1 (VLT, Paranal). Our optical identification of the bright submm source, which we henceforth refer to as SMMJ04135+1027, is an $I = 20.5 \pm 0.1$ mag point source. There are no obvious optical counterparts for the faint submm source. In Fig. 1 the I–band image is shown overlayed by the SCUBA contours.

The Galactic extinction of A478 is $E(B - V) = 0.52$ mag (Schlegel et al 1998). Hence, the corrected I magnitude is 19.5 ± 0.1 mag assuming a Milky Way type extinction law with $R_V = 3.1$. All the optical observations presented in the rest of this paper have been corrected accordingly.

To determine the redshift and the nature of SMMJ04135+1027 we have obtained an optical spectrum also with FORS1. The spectrum was observed in MOS mode with resolution $R = 150$. The spectrum exhibits the characteristics of high redshift quasars, such as broad emission lines (e.g. Lyα, CIV, CIII and SiIV), Lyman Forest absorption bluewards of the Lyα emission line, and a power law continuum. The redshift has been measured using the CIII line to be 2.83 ± 0.01, which is consistent with the other emission lines. The spectrum

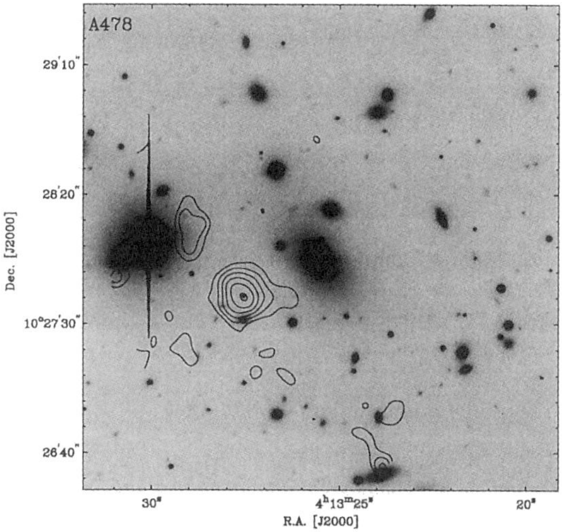

Figure 1: The I–image of A478 overlayed with the contours of the SCUBA 850 μm data. The contours each represent a step of 4 mJy. SMMJ04135+1027 is the very bright SCUBA source located SE of the center of the cluster. The other fainter possible submm source is located NE of the quasar and nearby the very bright star.

is shown in Fig. 2. If omitting the correction for Galactic extinction, the quasar appears very red. When corrected, as mentioned above, the quasar has a colour similar to optically selected quasars (Francis et al. 1992).

3 Discussion

3.1 Optical spectrum

The spectrum shown in Fig. 2 shows a number of unusual features when compared to the QSO sample of Francis et al. (1992). In the first place, Lyα is strongly suppressed. The origin of the suppression may be continuum absorption by dust, or Lyα absorption associated with the QSO. A higher resolution spectrum is needed to distinguish between these possibilities. Secondly, the CIV emission line is remarkably strong with an equivalent width of 110 Å.

3.2 The Far–IR SED

The fluxes quoted here have not been corrected for the weak magnification arising from the gravitational lensing caused by the intervening galaxy cluster.

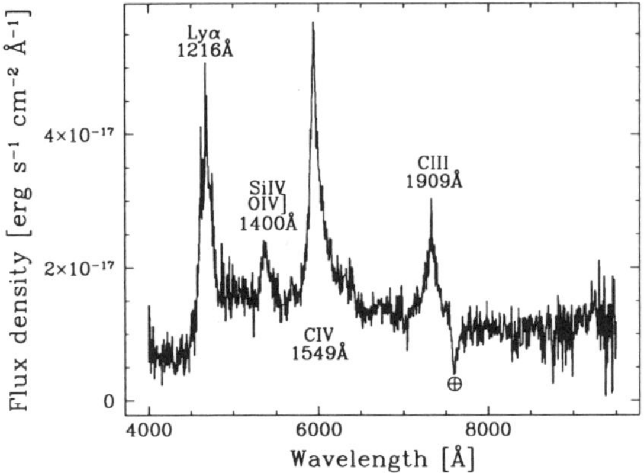

Figure 2: The spectrum of SMMJ04135+1027, corrected for Galactic extinction.

In this case, no arcs are detected in the vicinity of the quasar, hence the gravitational magnification is expected to be small and achromatic.

We have plotted (Fig. 3) the two SCUBA data points together with four known far–IR SEDs (Klaas et al 1999, Haas et al 1998, Sodroski et al 1997). The four SEDs have been redshifted to the redshift of the quasar and scaled to the quasar flux at $\lambda_{obs} = 850$ μm. We see that the data points correspond well to the SEDs of the two starburst galaxies, Arp220 and the Antennae, and to the low-z QSO PG0050+124. The cool Milky Way type of dust can easily be ruled out. Shorter wavelength data would be needed to see if the hot dust component that is found in low–z QSOs such as PG0050+124 is also present in SMMJ04135+1027. For an Arp220 SED, the total luminosity of SMMJ04135+1027 is $3 \times 10^{13} L_\odot$.

3.3 Contribution of AGNs to the submm samples

A number of well-studied submm sources exhibit AGN features in their optical spectra (e.g. Ivison et al. 1998). These are also typically the brightest submm sources in the samples. However, the fraction of AGNs in submm samples is still a debated issue. Optical spectroscopy of the fainter submm sources has turned out to be very challenging, therefore it will be difficult to fully characterize the submm population with optical spectroscopy alone. As shown by Fig. 3, shorter wavelength IR data may be used to find QSO–like dust emission. The most promising method for finding AGNs in submm samples is by hard

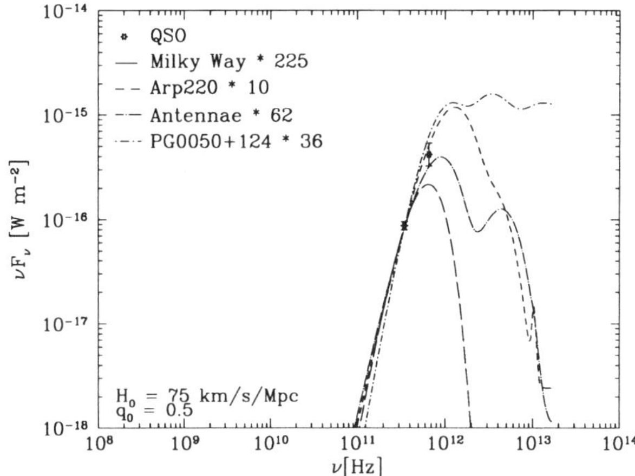

Figure 3: The 850 μm and 450 μm data points overlayed with the redshifted far–IR SEDs of the Milky Way, Arp220, the Antennae and PG0050+124.

X–ray observations. A recent study with *Chandra* and SCUBA of the relation between the resolved X-ray background and the resolved FIR background finds almost no overlap between the two populations, which suggests that SCUBA sources primarily are powered by starburst (Fabian et al. 2000). This result then suggests that the fainter submm sources contain relatively few AGNs, and that the AGNs that *are* present in the submm samples are to be found at the bright submm flux levels.

1. Fabian et al., 2000, MNRAS, 315, L8
2. Francis et al., 1992, ApJ, 398, 476
3. Haas et al., 1998, ApJ, 503, L109
4. Ivison et al., 1998, MNRAS, 298, 583
5. Klaas et al., 1999, A&A, 325, L21
6. Sodroski et al., 1997, ApJ, 480, 173
7. Schlegel et al., 1998, ApJ, 500, 525

8. The Future of Deep Sub-mm and mm Surveys

SIMULATED SUBMILLIMETRE GALAXY SURVEYS

DAVID H. HUGHES & ENRIQUE GAZTAÑAGA

Instituto Nacional de Astrofísica, Óptica y Electrónica (INAOE),
Luis Enrique Erro 1, Tonantzintla, Cholula, 78840 Puebla, Mexico

Current submillimetre (sub-mm) surveys are hindered in their ability to reveal detailed information on the epoch of galaxy formation and the evolutionary history of a high-redshift starburst galaxy population. The difficulties are due to the small primary apertures (D < 15-m) of existing sub-mm telescopes and the limited sensitivities of their first generation of bolometer cameras. This situation is changing rapidly due to a variety of powerful new ground-based, airborne and satellite FIR to mm wavelength facilities. Improving our understanding of the luminosity and clustering evolution provides the motivation for conducting cosmological sub-mm and mm surveys. It is therefore important that we quantify the limitations of the future surveys and the significance of the results that can be drawn from them. In this paper we present realistic simulated surveys which provide the means to address some key issues confronting existing and forthcoming surveys. We discuss the results from simulations with a range of wavelengths (200μm – 1.1 mm), spatial resolutions (6 – 27 arcsec) and flux densities (0.01–310 mJy). We describe how the measured source-counts could be affected by resolution and confusion, by the survey sensitivity and noise, and by the sampling variance due to clustering and shot-noise.

1 Introduction

By early 2001 the initial extensive programme of extragalactic SCUBA (850μm) surveys conducted on the 15-m JCMT will be completed. The targets of these sub-mm surveys include lensing clusters[16], the Hubble Deep Field[10], the Hawaii Deep Survey fields[1,2], the Canada-France Redshift Survey fields[5,14] and the UK 8 mJy SCUBA surveys of the ISOPHOT ELAIS and Lockman Hole field[3]. These blank-field sub-mm surveys, which cover areas of $\sim 0.002 - 0.2\,\mathrm{deg}^2$, allow a measure of the 850μm source-counts between flux densities of 1–15mJy. The relatively small survey areas are due to the low mapping speed (a combination of the field-of-view and sensitivity of the bolometer array, and the necessary overheads to fully sample the focal-plane with $2F\lambda$ feed-horns. The restricted 850μm flux range in the measured source-counts is the result of a source-confusion limit of $\sim 2\,\mathrm{mJy}$ (due to a small primary aperture or map resolution) at faint flux densities and the small survey area at the bright end, given the intrinsic low source-density of bright sub-mm sources (*e.g.* $N[S_{850_{\mu m}} > 20\,\mathrm{mJy}] < 70$ sources/deg^2).

The limitations on the accuracy of our understanding of high-z galaxy evolution from sub-mm surveys has been described elsewhere[11]. To improve

Table 1. Number (N) of 5-σ galaxies detected in alternative 50-hour BOLOCAM-2 1.1 mm surveys during routine operation on the LMT. A conservative Noise Equivalent Flux Density (NEFD) of 4 mJy sec$^{1/2}$ at 1.1 mm is adopted assuming a primary aperture r.m.s. surface accuracy of $\eta \simeq 70\mu m$. Surveys A and B represent LMT surveys similar in area to the HDF-SCUBA survey[10] and the UK SCUBA 8 mJy wide-area survey[3] respectively. Surveys C and D represent future wide-area LMT surveys.

Survey	Area (sq. arcmin)	3-σ depth	N (5-σ galaxies)
LMT-A	6	0.05 mJy	33
LMT-B	350	0.4 mJy	515
LMT-C	1000	0.6 mJy	1100
LMT-D	3600	1.2 mJy	1600

the constraints on the competing evolutionary models, future sub-mm and mm surveys must extend their wavelength coverage, increase their survey area and sensitivity. To quantify the advantages of these forthcoming surveys we have simulated the extragalactic sky at various sub-mm and mm wavelengths (200-3000μm), spatial resolutions ($\theta_{FWHM} \sim 1 - 120$ arcsec) and sensitivities ($S > 0.01$ mJy). We include contributions from CMB primary fluctuations, S-Z clusters, an evolving extragalactic population of starburst galaxies and high-latitude galactic-plane cirrus emission. These simulated surveys allow us to consider the observing strategies and extragalactic science for a wide variety of future FIR and sub-mm/mm telescopes (*e.g.* SIRTF, FIRST, MAP, PLANCK, BLAST, SOFIA, LMT, GBT, ALMA). The details of these multi-wavelength simulations are discussed elsewhere[12].

In this paper we concentrate on a discussion of the distortion of the measured source-counts due to galaxy clustering and low spatial resolution. We illustrate the issues with results from simulated surveys at 1.1 mm, 850μm, 350μm and 200μm with spatial resolutions of 6, 15, 9 and 25 arcsec corresponding to surveys on the 50-m LMT (http://www.lmtgtm.org), 15-m JCMT, 10-m CSO and 2-m BLAST (http://www.hep.upenn.edu/blast/) respectively. The full simulations have a flux dynamic range of $S_\lambda = 0.01 - 310$ mJy at all wavelengths and cover an area of 1 sq. degree. Subsets have been extracted to determine the source-counts from surveys comparable in area (6–400 arcmin^2, Table1 - survey A and B) to the SCUBA surveys of the Hubble Deep Field and Hawaii Deep Fields, the lensing cluster survey, Canada-France Redshift survey fields and the UK 8 mJy ELAIS and Lockman Hole survey. Additionally larger-area surveys (0.3–1.0 deg^2, Table1 - surveys C and D) more appropriate to the LMT and BLAST are also considered.

Fig.1 shows a comparison of the same 0.1 deg^2 map, selected from a

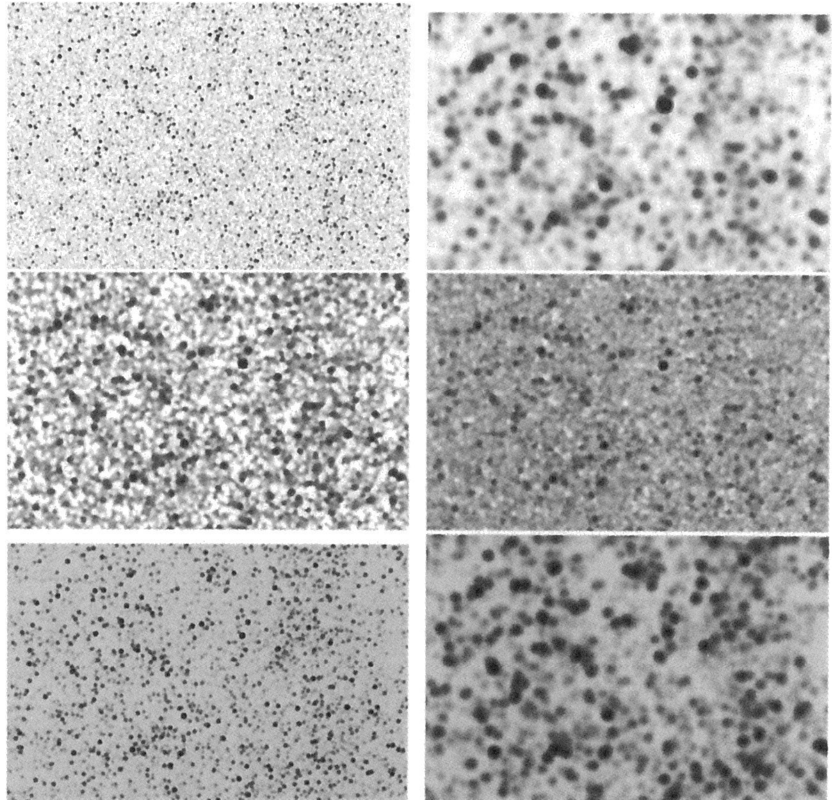

Figure 1. Simulated surveys covering $0.1\,\mathrm{deg}^2$: A confusion-limited ($3\sigma = 0.05\,\mathrm{mJy}$, $\theta_{FWHM} = 6$ arcsec) LMT survey at 1100μm (top-left); a $\theta_{FWHM} = 27$ arcsec low resolution 1100μm survey (top-right); a $\theta_{FWHM} = 15$ arcsec deep ($1\sigma = 0.5\,\mathrm{mJy}$) 850μm survey (middle-left); a $\theta_{FWHM} = 15$ arcsec shallow ($1\sigma = 2.5\,\mathrm{mJy}$) 850μm survey (middle-right); a confusion-limited ($\theta_{FWHM} = 9$ arcsec) 350μm survey (bottom-left); a confusion-limited ($\theta_{FWHM} = 25$ arcsec) 200μm survey (bottom-right).

$1.0\,\mathrm{deg}^2$ simulation, at different resolutions and wavelengths. In general, one can recognise the brightest sources in an individual map at any other wavelength. Consequently it is possible to derive a robust constraint on the redshift of an individual galaxy from the relative intensities at different wavelengths[11]. However care must be taken in determining the appropriate sensitivities of the complementary surveys when the primary goal is to measure mm to sub-mm colours to estimate the redshift distribution of the entire population[13].

2 Source-Count Analysis

A comparison of the expected number of galaxies at 850μm and 200μm in a given log redshift bin, $N(z)$, as given by the input selection function, and the measured counts in the redshift simulations (before including the noise or the finite resolution to the map) demonstrates that the simulations accurately reproduce the input models.

A potential problem arises when a single representative galaxy SED is used to calculate the luminosity as a function of redshift, and subsequently to predict the source-counts. However since the dispersion in the rest-frame SEDs of galaxies is greatest in the mid-IR to FIR regime (at $\lambda > 60\mu$m), the choice and/or evolution of the galaxy SED will only affect the counts at the shortest sub-mm wavelengths, $< 250\mu$m, and then only for those galaxies at the highest redshifts, $z > 3$.

The number counts and photometry of the sources in our simulated angular maps were determined using SEXTRACTOR (within the STARLINK package GAIA). In Fig.2 we compare the predicted and measured number-counts of the sources extracted from the full 1-deg^2 simulations at 1100, 850 and 200μm. Again there is excellent agreement between the simulated source-counts and the model down to the confusion flux limit of the different surveys at their respective resolutions. This illustrates the obvious point that provided the survey is of sufficient area and sensitivity then neither resolution, projection or noise are very important in extracting counts for objects above the confusion-limit.

2.1 Resolution and confusion

Below the confusion limit the counts flatten as faint sources merge to form brighter objects (see Fig.2). For example, at 850μm (with 15 arcsec resolution) the counts are affected by confusion at $S_{850\mu m} \leq 2\,$mJy, whilst at 1.1 mm with 6 arcsec resolution (*e.g.* LMT) the counts are still unaffected at $S_{1.1mm} \sim 0.1\,$mJy. Even at lower resolution (27 arcsec) the measured 1.1 mm source-counts recover the input model down to $S_{1.1mm} \sim 3\,$mJy due to the low density of intrinsically luminous sources ($> 10^{12}L_\odot$). Similarly the 200μm counts recover the model down to a confusion limit of 18 mJy at 25 arcsec resolution. Hence even at resolutions of $\sim 30\,$arcsec the overlapping of individual source PSFs is negligible, regardless of whether it's due to the random line-of-sight projection of galaxies at different redshifts, or because of a high-amplitude of clustering. The significance of this result is that it is therefore possible to conduct bright surveys, combining large-aperture (50-m LMT, 15-m JCMT)

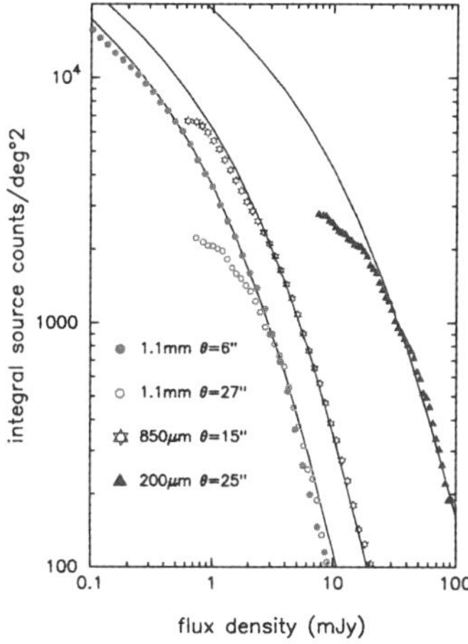 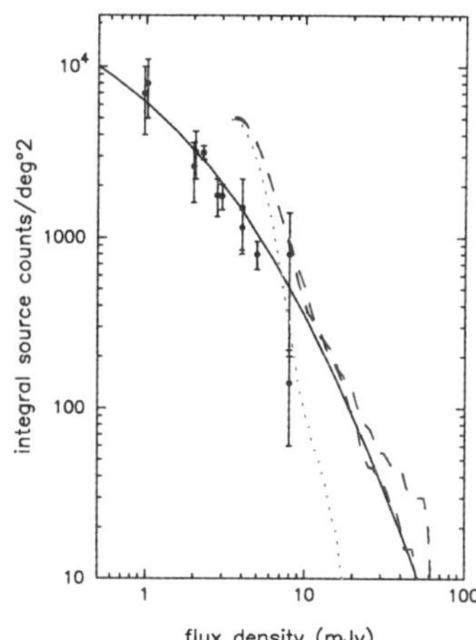

Figure 2. Left-hand figure: A comparison of the input model (solid-lines) and the extracted source-counts from the simulated 1-deg^2 surveys at 200μm, 850μm and 1.1 mm. Note how the simulated counts are depleted with respect to the input model at fluxes fainter than the confusion limit, which is set by the spatial resolution of the survey. The 1.1 mm counts at different resolutions (6 arcsec and 27 arcsec) are also shown. Right-hand figure: Extracted number-counts at 850μm from two ($3\sigma \sim 8$ mJy) 0.2 deg^2 simulated surveys (dashed-lines). The symbols show the actual observed SCUBA 850μm counts from the surveys described in §1 and the solid line corresponds to the evolutionary model which fits the observed counts (which is the input to the simulations). The counts from the noise map ($1\sigma = 2.5$ mJy), which was added to the raw simulation, are shown as a dotted line. Because these are not deep surveys, the noise dominates the counts producing spurious simulated sources at 8-12 mJy. This has the opposite effect to the confusion produced by the finite resolution.

3000–450μm wavelength surveys and relatively small primary aperture (*e.g.* ~ 2.0 m BLAST) sub-mm and FIR surveys (500 − 200μm) with very different resolutions ($\theta_{FIR}/\theta_{mm} \sim 5$), and still derive meaningful colours of luminous sub-mm-selected galaxies, and hence constrain their redshifts[11].

2.2 Extragalactic Background

The 1.1 mm extragalactic background with flux $S > 0.5$ mJy due to the discrete sources in our simulations is 1.31×10^{-10} Wm^{-2}sr^{-1}, in excellent agreement with FIRAS (Far Infrared Absolute Spectrophotometer) results from COBE residual measurement of the extragalactic far infrared background[4]. This implies that surveys conducted at the full spatial resolution of the LMT should resolve $\sim 100\%$ of the mm background, whilst the deepest SCUBA surveys to-date have resolved only 30-50% of the sub-mm background.

2.3   Depth and noise

In the previous subsection we indicated that provided surveys have sufficient sensitivity and reach their confusion limit then accurate source-counts can be measured from large-area surveys. However the simulations also illustrate the way in which the counts are severely affected for shallow wide-area surveys with a flux-limit well above the confusion limit when the number-counts are steep. For example Fig.2b shows the simulated counts from a $0.2\,\mathrm{deg}^2$ 850μm SCUBA survey with a 1-σ sensitivity of 2.5 mJy. The extracted counts accurately reproduce the model down to a flux density of ~ 12 mJy, below which a steep increase in the counts is observed due entirely to the noisy background. Hence individual sources identified in shallow sub-mm surveys at the 3–5σ level may be spurious. These additional sources can have a non-negligible positive contribution to the overall number counts (compared to the situation in §2.1 where the counts flatten at the confusion-limit of the survey). This is due to an increasing contribution of the noise counts, relative to the measured galaxy counts (which steepen as the surveys move to brighter limiting flux densities). However as the noise spectrum is well determined (*e.g.* dashed-line in Fig.2b) a correction can be applied to remove the excess counts[12].

2.4   Clustering and shot-noise

Another important aspect illustrated by the simulations is the effect induced on the counts by the sampling variance of the large-scale galaxy clustering. This is most significant in the deepest surveys which necessarily provide the smallest survey areas ($< 0.1\,\mathrm{deg}^2$). To quantify this effect, three deep surveys, identical in area ($\sim 6\,\mathrm{arcmin}^2$) to the SCUBA survey of the Hubble Deep Fields[10] and similar to those of the Hawaii Deep Fields[1] and lensing galaxy clusters[16], were randomly selected from the full simulations. We find a factor of 3–10 variation in the extracted counts between these deep confusion-limited surveys. The three simulated deep surveys and the original SCUBA HDF survey are shown in Fig.3. For the SCUBA surveys we would need an area over 100 times larger if we want to reduce the variance in the counts between different selected regions to the few percent level. Note that besides the intrinsic clustering, shot-noise plays an important role in this variance, at least for the brighter (less numerous) sources. A detailed analysis of all these effects are presented elsewhere[12].

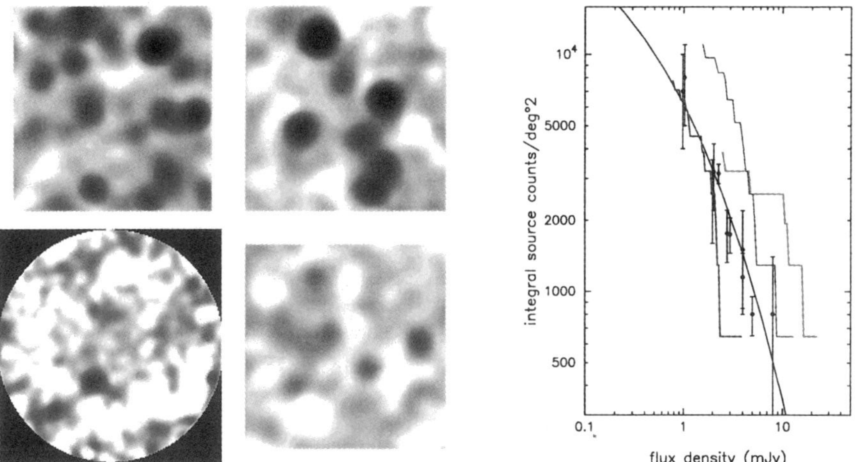

Figure 3. Left-hand figure: Three deep $850\mu m$ surveys extracted from the full 1 sq. deg^2 simulation with the same area and depth as the SCUBA $850\mu m$ survey of the HDF[10] (shown as a circle inside a black square for comparison). Right-hand figure: Histograms representing the $850\mu m$ source-counts for the three different small area surveys ($\sim$ 6 arcmin^2). The solid-line represents the model, fitted to the measured $850\mu m$ counts, that produces the full simulation.

3 Conclusions

We have presented realistic extragalactic simulations which can address some key issues confronting existing and forthcoming mm and sub-mm surveys. For example, what is the effect of the wavelength-dependent resolution and confusion on the extracted mm and sub-mm colours of individual sources, which provide crude, but essential photometric redshift information[13]? Equally important, how does the survey sensitivity and sampling variance due to clustering and shot-noise affect the measured source-counts at the faintest and brightest flux levels where the data have the most diagnostic power to discriminate between competing evolutionary models? Both the number counts and the two-point clustering pattern depend strongly on the cosmological model, but its combination can break some degeneracies. If we consider a scenario where galaxy formation occurs in the rare peaks of the underlying matter field then the resulting galaxy clustering should be stronger than if galaxy formation occurred in random places. The measure of the coherence of the clustering pattern, as described by the higher-order statistics (which is apparent in our simulated maps), can be used to test how star forming galaxies trace the underlying mass[8]. These issues are addressed in detail in a full paper[12].

References

1. Barger, A.J. *et al*, *Nature* **394**, 248 (1998).
2. Barger, A.J., Cowie, L.L., Sanders, D.B., *ApJ* **518**, L5 (1999).
3. Dunlop, J.S., this volume.
4. Dwek, E. *et al*, *ApJ* **508**, 106 (1998).
5. Eales, S.A. *et al*, *ApJ* **515**, 518 (1999).
6. Gaztañaga, E., Baugh, C., *MNRAS* **294**, 229 (1998).
7. Gaztañaga, E., *ApJ* **454**, 561 (1995).
8. Gaztañaga, E., astro-h/0003160, (2000).
9. Giavalisco, M. *et al*, *ApJ* **503**, 543 (1998).
10. Hughes, D.H. *et al*, *Nature* **394**, 241 (1998).
11. Hughes, D.H., astro-ph/0003414, (2000).
12. Hughes, D.H., Gaztañaga, E., *MNRAS*, submitted (2001).
13. Hughes, D.H., Gaztañaga, E., Aretxaga, I., Chapin, E., Dunlop, J.S., *MNRAS*, submitted (2001).
14. Lilly, S. *et al*, *ApJ* **518**, 441 (1999).
15. Saunders, W. *et al*, *MNRAS* **242**, 318 (1990).
16. Smail, I. *et al*, *MNRAS* **490**, L5 (1997).

SZ EFFECT IN YOUNG MASSIVE ELLIPTICALS?

DANIEL ROSA GONZÁLEZ[1*], ROBERTO J. TERLEVICH[2], ELENA
TERLEVICH[1], AMANCIO FRIAÇA[3], ENRIQUE GAZTAÑAGA[1]

[1] *Instituto Nacional de Astrofísica, Optica y Electrónica, INAOE, Luis Enrique
Erro 1, Tonantzintla, Puebla 72840. México.*
[2] *Institute of Astronomy, Madingley Road, CB3 OHA Cambridge, U.K.*
[3] *Instituto Astronomico e Geofísico, USP, Av. Miguel Stefano 4200, 04301-904 Sao
Paulo, SP, Brazil*

We study the Sunyaev-Zel'dovich (SZ) effect during the formation of elliptical
galaxies, and estimate the epoch where the observation of giant elliptical galaxies
through the SZ effect would be more efficient. The elliptical galaxy formation
models of Friaça & Terlevich[1] are used. We show that during the early epochs of
galaxy formation ($\sim 3\times10^8$ years) the gas in the center of giant elliptical galaxies
($\sim10^{12}M_\odot$) reaches a temperature and density high enough to detect the SZ
signature in reasonably short exposure times using a big collecting area telescope
equipped with a sensitive bolometer millimeter array (e.g. BOLOCAM in the
future LMT/GTM).

1 Introduction

The SZ effect is the distortion of the Cosmic Microwave Background (CMB)
due to the inverse Compton scattering produced when the radiation goes
through the hot plasma confined in the gravitational potential of a cluster[2].
In recent years, after the first measurements of H_0 by combining X-rays with
the SZ effect, groups have begun to report the measurement of the SZ effect in
a significant number of clusters using single dish radiometric, bolometric and
interferometric observations (e.g. McHardy et al.[3] on Abell 2218 or Birkinshaw, Hughes & Arnaud [4] on Abell 665, see also Birkinshaw [5] for an extensive
and recent review). Using models for galaxy formation we discuss the possibility of observing the SZ effect in giant elliptical galaxies. In fact, the gas in
these galaxies during the early stages of evolution could reach temperatures
and densities high enough to get a Comptonization parameter comparable
to that observed in galaxy clusters. We show that in the case of very massive elliptical galaxies it may be possible to detect the SZ signature using a
big collecting area telescope (like the GTM/LMT 50 meter telescope) and a
millimeter array such as BOLOCAM.

---

*E-MAIL: DANROSA@INAOEP.MX

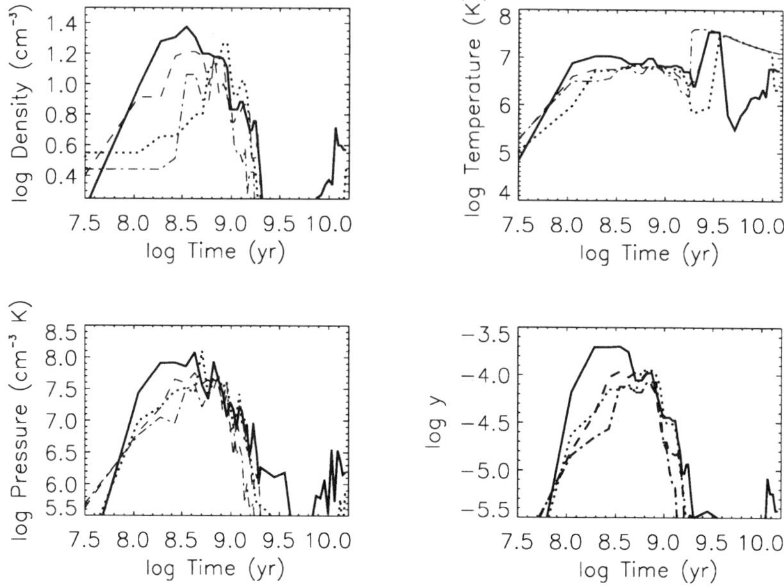

Figure 1. Density, temperature, pressure and y parameter as a function of time. The baryonic masses are of 2×10^{12} solar masses for the solid line, 10^{12} for the dotted line, 5×10^{11} for the dashed line and 2×10^{11} for the dot-dashed line. The observed flux can be estimated from the lower right panel taking into account that for the LMT/GTM at 1 mm a y parameter of 10^{-4} corresponds to 1 mJy. y corresponds to the inner 7 kpc.

2   The Elliptical Galaxy Formation Model

The Friaça and Terlevich models of galaxy formation combine multi-zone chemical evolution with 1-D hydrodynamics. The hydrodynamical evolution is given by the resolution of the fluid equations of mass, momentum and energy conservation plus an equation of state. The total mass of the elliptical galaxy is given by the contribution of the stellar mass, the gas and the dark halo. The chemical evolution is driven by the stellar winds, planetary nebulae and SN phase which produce the enrichment of the surrounding ISM. Outputs of the models are the temperature and density of the gas as a function of time (shown in Figure 1) and distance to the center.

3 The SZ Effect in Giant Elliptical Galaxies

The SZ effect can be parameterized using the Comptonization parameter y defined by the integral of the electron pressure along the line of sight.

$$y \propto \int dl \; n_e \; T_e \tag{1}$$

In this equation n_e and T_e are the electron number density and the electron temperature respectively. The SZ spectrum is calculated using the non relativistic expression given by the Kompaneets approximation[6].

The SZ effect has been observed in galaxy clusters with typical radius of 100 kpc, temperatures of about 8 keV ($\sim$ 115 x 10^6K) and electron densities of about 10^{-2} cm^{-3}. These values give a central y of about 10^{-4}. In Figure 1 we show that massive galaxies with ages between 10^8 and 10^9 years reach a maximum central temperature of about 10^7 K and a density of about 5 cm^{-3}. These values of temperature and density give a y larger than 10^{-4} during this period which is comparable to those values of y calculated for galaxy clusters. The y parameter was calculated by integrating across the 7 central kiloparsecs. In this region the temperature and density of the gas remain almost constant.

In Figure 2 (left) we show the expected flux in the case of a 50 meters telescope operating at 1 mm. Note that the signature of the SZ effect is independent of redshift, so as long as the galaxy is resolved the SZ effect can be measured towards arbitrarily distant galaxies by taking into account the corresponding dilution effect. As shown in the figure, with a 1-hour integration with BOLOCAM, it is possible to detect at the 3 σ level fluxes larger than 0.1 mJy, which corresponds to galaxies with M $\geq 10^{12}$ solar masses (for any value of $0 < \Omega_\Lambda < 1.5$ and $\Omega_M = 0.3$).

We used the Press Schechter formalism (PS) to estimate the number of halos of a given mass as a function of the redshift[7]. The bias parameter was fixed to unity ($\sigma_8 = 1$) and the power of the initial mass spectrum was calculated for each mass range assuming a ΛCDM cosmology[8]. The results are shown in Figure 2 (right) for dark matter masses of (4 ± 1)x10^{11} and (2 ± 0.5)x10^{13}, which correspond to the baryonic masses in the left panel. The labels $N = 3.3 \times 10^5$ and $N = 3.2 \times 10^3$ in each case indicate the total number of halos of that mass integrating from $z = 0$ to $z = 50$. This is the number of massive halos that we could potentially detect in a 1 square degree field with a depth given by the fluxes in the left panel for each case. In practice we need to correct for other sources of mm emission (like dust) and take into account the efficiency of conversion of halos into elliptical cores.

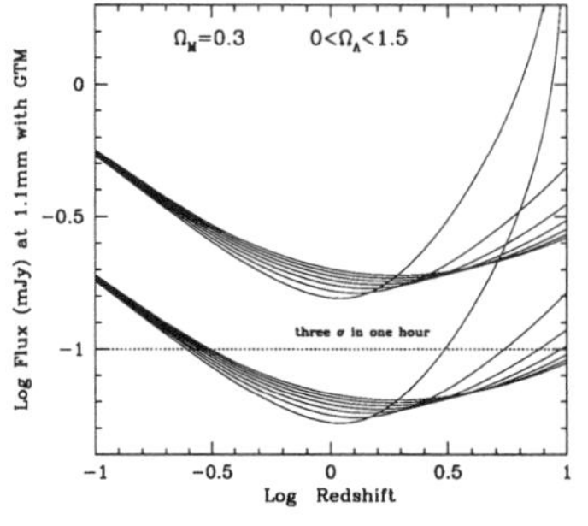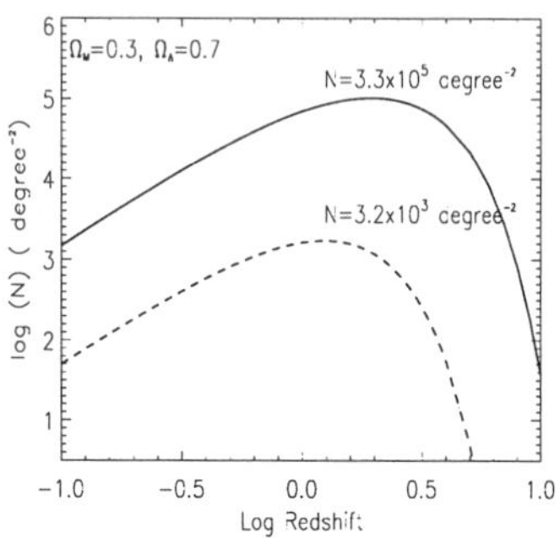

Figure 2. **Left panel:** Dilution effect due to the change in the angular size with redshift. The predicted rms were calculated assuming a sensitivity of 2 mJy/$\sqrt{Hz}$. The lower (upper) family of curves corresponds to the model of 10^{11} (5×10^{12}) solar masses in baryonic matter. **Right panel:** Number of dark halos per unit redshift per square degree as a function of look back time using the PS formalism for two different mass ranges. The dark matter masses are $(4\pm1)\times10^{11}$ and $(2\pm0.5)\times10^{13}$ for the solid and dashed lines.

4 Conclusions

With the advent of LMT/GTM and ALMA and their high sensitivity and resolution it may be possible to detect the SZ effect in individual young ellipticals that although cooler, smaller and denser than clusters, may have similar Comptonization parameter in their central regions. Using the PS formalism we show that the number of massive halos that could produce a detectable SZ signature is significant.

References

1. A.C.S. Friaça, and R.J. Terlevich, MNRAS **298**, 399 (1998)
2. R.A. Sunyaev, and Y.B. Zel'dovich, Comm. Astrophys. and Sp. Phys. **4**, 173 (1972)
3. I.M. McHardy, G.C. Stewart, A.C. Edge, B. Cooke, K. Yamashita, and I. Hatsukade, MNRAS **242**, 215 (1990)
4. M. Birkinshaw, J.P. Hughes, and K.A. Arnaud, ApJ **379**, 466 (1991)
5. M. Birkinshaw, Phys. Rept. **310**, 97 (1999)
6. A.S. Kompaneets, Zh.E.F.T. **31**, 876 (1957)
7. W.H. Press, and P. Schechter, ApJ **187**, 425 (1974)
8. P.T.P. Viana, and A.R. Liddle, MNRAS **303**, 535 (1999)

FIR/SUB-MM LINE EMISSION FROM THE FIRST OBJECTS: TESTING THE STELLAR FEEDBACK

B. CIARDI

Universitá di Firenze, Largo Enrico Fermi 5,
50125 Firenze, Italy
E-mail: ciardi@arcetri.astro.it

A. FERRARA

Osservatorio Astrofisico di Arcetri, Largo Enrico Fermi 5,
50125 Firenze, Italy
E-mail: ferrara@arcetri.astro.it

We calculate the expected FIR/sub-mm molecular hydrogen-line emission from the first galactic objects that formed in the universe. Due to their low masses, the stellar feedback from massive stars is able to blow away their gas content and collect it into a cooling shell where H_2 rapidly forms and IR rotational lines carry away a large fraction of the explosion energy. The expected fluxes from these sources are in the range 10-1000 μJy. The highest number counts are expected in the 60-80 μm band. Among the planned FIR/sub-mm facilities, we find that the best detection opportunity is with the *Multiband Imaging Photometer* on SIRTF.

1 Introduction

Detecting the first objects that formed in the Universe will be the primary goal of several future space/ground-based telescopes. These low mass systems are strongly affected by feedback mechanisms both of radiative and stellar type[2]. Molecular hydrogen, being the only available coolant in a plasma of nearly primordial composition, is a key species in the feedback network as it regulates the collapse and star formation in these objects. Ferrara[3] has pointed out that H_2 is efficiently formed in cooling gas behind shocks produced during the blow-away (*i.e.* the complete ejection of the galactic gas) process thought to occur in the first objects, with typical fractions of $f_{H_2} \approx 6 \times 10^{-3}$. The conditions in these cooling blastwaves are such that a noticeable amount of the explosion energy is carried away by infrared H_2 molecular lines, which therefore might provide us with a superb tool to detect these primordial galactic blocks.

2 Emission model

The first detectable H_2 emission lines are produced by quadrupole radiation and they are purely rotational. The lines relevant to the present work are listed

Table 1. Relevant H_2 Transition Lines

Transition	$\Lambda^{(1)}$ [μm]	$T^{(2)}$ [K]	$A^{(3)}$ [s^{-1}]
0-0S(0)	28.0	512	2.94×10^{-11}
0-0S(1)	17.0	1015	4.76×10^{-10}
0-0S(2)	12.3	1681	2.76×10^{-9}
0-0S(3)	9.7	2503	9.84×10^{-9}
0-0S(5)	6.9	4586	5.88×10^{-8}

[1]Emission wavelength; [2]Excitation temperature; [3]De-excitation Einstein coefficient.

in Table 1. The temperatures required to excite these lines, 500–4500 K, are reached during the cooling of the post-shock IGM gas produced during the blow-away of low-mass primordial galaxies [3]. It is then conceivable to expect that the above molecular lines are excited during this process, producing potentially detectable radiation. The range of masses that can experience a blow-away and produce such radiation is shown in Fig. 1. The predicted flux on the ground in a given line is derived as follows. Let ν_{ik} be the rest-frame frequency of photons emitted by the molecule during the transition between the energy levels k and i. The line emissivity is then given by:

$$j_{\nu_{ik}} = \frac{h}{4\pi} \nu_{ik} n_k A_{ki} \Phi(\nu_{ik}), \tag{1}$$

where n_k is the number density of molecules in the k level, A_{ki} is the Einstein coefficient for spontaneous emission and $\Phi(\nu_{ik})$ is the line profile function calculated at the line center [5]. As the conditions for thermodynamic equilibrium are not satisfied, the population of the levels must be obtained by solving the detailed balance equations[1]. Following from eq.1, the observed flux is:

$$F(\nu_o) = L_{\nu_{ik}}(1 + z)/4\pi d_L^2 \tag{2}$$

where $\nu_o = \nu_{ik}/(1 + z)$ is the observed frequency and d_L is the cosmological luminosity distance, derived for a ΛCDM cosmology with $\Omega_M = 0.35$, $\Omega_\Lambda = 0.65$, $\Omega_b = 0.04$ and $h = 0.65$. $L_{\nu_{ik}} = 4\pi V j_{\nu_{ik}}$ is the specific luminosity, where $V = (4/3)\pi R_s^2 dR_s$ is the physical volume occupied by the H_2 forming shell. The values of R_s and dR_s are obtained from the formulae in Ferrara (1998) and an explicit expression is found in Ciardi & Ferrara (2000).

Additionally, one can derive the number of objects whose observed flux is larger than the threshold value F_{min}, in a redshift interval $z_{min} < z < z_{max}$; the limits depend on the observational wavelength band of the instrument.

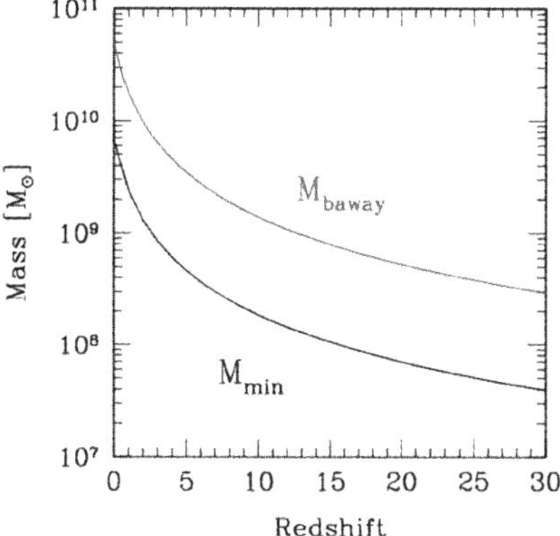

Figure 1. Maximum mass for an object to be able to experience a blow-away (upper line) and minimum mass present (lower line) as a function of redshift[2].

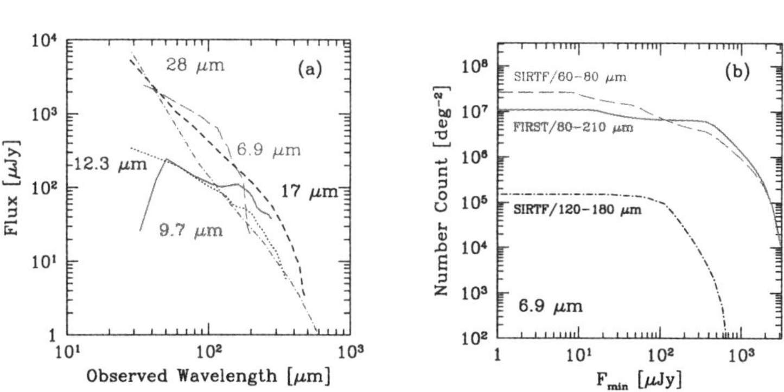

Figure 2. (a) Expected flux as a function of the observed wavelength for the five H_2 molecular lines considered: 6.9 μm (long-dashed line), 9.7 μm (solid), 12.3 μm (dotted), 17 μm (short-dashed) and 28 μm (dot-dashed). The flux is derived for the average halo mass evaluated at the emission redshift. (b) Number counts from the H_2 6.9 μm emission wavelength as a function of the minimum observed flux, F_{min}. The number counts are calculated for different observed wavelength bands: 60-80 μm (long-dashed line), 80-210 μm (solid) and 120-180 μm (dot-dashed).

3 Results

Fig. 2a shows the expected flux as a function of the observed wavelength, λ_{obs}, in the five different lines 6.9 μm, 9.7 μm, 12.3 μm, 17 μm, and 28 μm. As an example, we plot the fluxes calculated for the mass of the halos that suffer blow-away averaged over the halo mass distribution, as given by the Press & Schechter [4] formalism, at the explosion redshift. In addition, the curves refer to emission occurring when the shocked gas temperature has decreased to the corresponding line excitation temperature. We see that the flux decreases with increasing wavelength, *i.e.* emission redshift.

The expected number of objects per unit sky area as a function of the limiting flux, F_{min}, of a given experiment is shown in Fig. 2b for the 6.9 μm line. The number counts have been estimated in three different FIR/sub-mm bands corresponding to the instrument passbands on board the planned space missions *Space InfraRed Telescope Facility* (SIRTF) and *Far InfraRed and Submillimeter Telescope* (FIRST). For the SIRTF 60-80 μm band, the number counts for the 6.9 μm line extend up to 2×10^3 μJy: at this flux level about 10^5 objects/deg^2 are predicted. A second promising band is at 80-210 μm (on FIRST) where the counts extend up to 3×10^3 μJy.

We now compare the above results with the forseen capabilities of SIRFT and FIRST; all sensitivities are given for a 3σ detection in one hour. The *Multiband Imaging Photometer* for SIRTF (MIPS) is expected to reach a sensitivity of $3 \times 10^2 \mu$Jy in the band 60-80 μm and of $5 \times 10^3 \mu$Jy in the band 120-180 μm. At the above sensitivity level, MIPS should be able to detect these primordial objects, although the the confusion limit of the instrument is not well determined. The lower MIPS sensitivity in the longer wavelength band 120-180 μm essentially prevents detections. The PACS photometer on board of FIRST (sensitivity range 80-210 μm) will reach a limiting flux of $1.8 \times 10^3 \mu$Jy, and therefore can only detect emission from the 6.9 μm line very marginally. In conclusion, SIRTF/MIPS appears to constitute the best instrument to reveal the emission from the first objects.

References

1. Ciardi, B., Ferrara, A., *MNRAS*, submitted, astro-ph/0005461 (2000).
2. Ciardi, B. *et al*, *MNRAS* **314**, 611 (2000).
3. Ferrara, A., *ApJ* **499**, 17L (1998).
4. Press, W.H., Schechter, P., *ApJ* **187**, 425 (1974).
5. Spitzer, L. Jr., Physical Processes in the Interstellar Medium, Wiley-Interscience, New York (1978).

NUMERICAL SIMULATION OF THE SUB-MM GALAXIES

MARK FARDAL, NEAL KATZ,
University of Massachusetts

ROMEEL DAVÉ,
University of Arizona

LARS HERNQUIST,
Harvard University

DAVID WEINBERG
Ohio State University

We study the sub-millimeter emission from massive galaxies in an N-body/hydro simulation. Assuming that most of the emission from newly formed stars is absorbed and reradiated in the rest infrared, we can calculate the expected number of galaxies detected in sub-mm surveys conducted with SCUBA. The number counts are strongly dependent on the assumed emission model. Assuming plausible values for the dust temperature and emissivity, we reproduce the counts at $\sim 3\,\mathrm{mJy}$, though the counts may be somewhat steeper than observed. The broad distribution of the simulated galaxies around a median redshift of $z \approx 2.5$ is in agreement with current observations. The simulated SCUBA sources are forming stars fairly steadily, and their properties agree with the notion that they are progenitors of massive ellipticals.

It is clear by now that sub-mm surveys select a population of galaxies that are forming stars rapidly at high redshifts. Modeling of the sub-mm galaxies has so far been done with phenomenological or semi-analytic methods.[1,2] Here we describe a numerical simulation that appears to generate the sub-mm population as a natural consequence.

The simulation is performed with a parallel version of the smoothed particle hydrodynamics code TreeSPH [3,4], which includes gas cooling and star formation. Simulations with this code have been successful in reproducing the clustering and evolution of galaxies[5,6], as well as the Lyman-α forest[7]. We assume $\Omega_M = 0.4$, $\Omega_\Lambda = 0.6$, $\Omega_b = 0.02h^{-2}$, and $H_0 = 65\,\mathrm{km\,s^{-1}\,Mpc^{-1}}$. We confine our study to a single run, with 144^3 particles in the gas and dark matter, and a box side of $50h^{-1}\,\mathrm{Mpc}$. This run is tuned to study the evolution of large $(L > 0.3L_*)$ galaxies down to $z = 0$.

We calculate the sub-mm fluxes in the simplest possible way. We identify the galaxies by grouping the baryonic particles, find their star formation rates, and convert these to bolometric luminosity using[9] $L_{bol} = 6.7 \times 10^9 L_\odot\, M_\odot^{-1}\,\mathrm{yr}$,

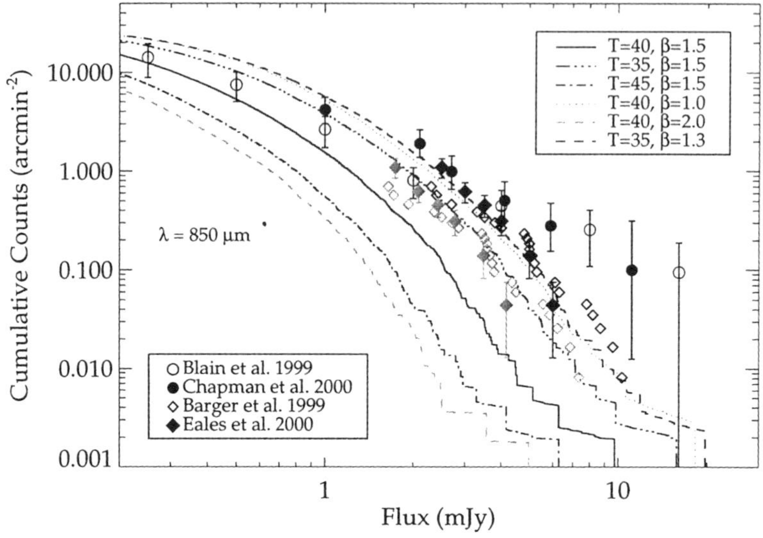

Figure 1. The counts as a function of flux at 850 μm. Simulation results are shown with various choices of the SED as indicated in the legend. The SED assumes a dust temperature T, an emissivity index β (so that the Planck function is multiplied by ν^{β}), and a short-wavelength tail so that f_{ν} falls no faster than ν^{-3}. Note that the two choices $T = 40K$, $\beta = 1.0$, and $T = 35K$, $\beta = 1.3$ are nearly coincident. Results of blank-field surveys[9,16] are shown as the diamonds, while results from lensing cluster surveys[10,11] are shown as circles. Lilly has argued at this conference that at least the blank-field counts should be shifted downward by 1.4 in flux to reflect the effect of noise and confusion; these are shown as the points in shadow.

consistent with our Miller-Scalo IMF. The resolution of the individual galaxies is low and metals are not included in the simulation, so we cannot compute *a priori* the fraction of this luminosity absorbed by dust. However, the faintness of the typical SCUBA source in the optical[8] suggests that these sources are in general highly obscured. In the figures here we assume that all of the energy comes out in the FIR.

To convert from FIR luminosity to 850 μm flux, we assume a fixed SED. Using discrete outputs from the simulation, we compute the 850 μm flux for each galaxy in each output, and replicate the computational box for each output over a shell in redshift corresponding to the output time. We can then measure the flux and redshift distributions. We do not consider AGN

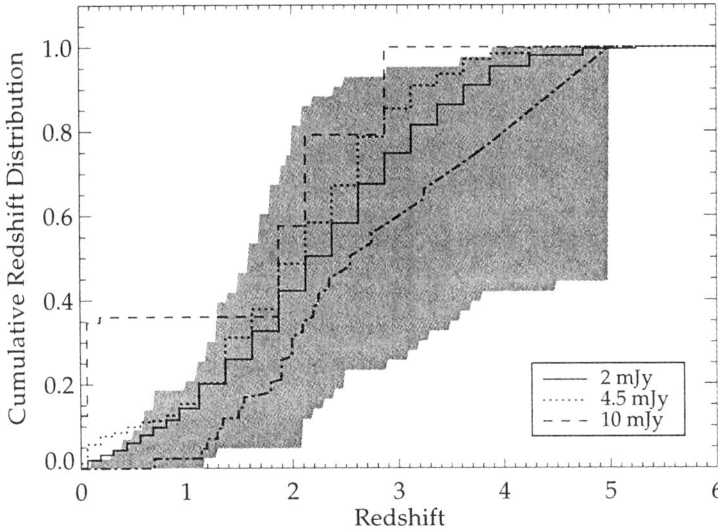

Figure 2. The cumulative redshift distribution for different flux-limited samples. The observational results[14,15,16] are highly uncertain, since the redshifts are mostly obtained from the radio-FIR correlation. The dot-dashed line shows the result from placing these sources at the middle of the redshift ranges given by the authors; the boundary of the gray shaded region results from placing all of them at the minimum or maximum of the redshift range.

emission, which will add to the counts, perhaps especially at the brighter end.

Figure 1 displays the counts computed from the simulation, and several observational samples. The computed counts are highly sensitive to the assumed SED. Our best fit uses $T = 35\,\mathrm{K}$, $\beta = 1.3$, which happen to be the mean values found in the SCUBA survey of local galaxies[12]. Adding random scatter to the SED parameters, comparable to that found in the local survey, would improve the fit further. The calculated counts may be somewhat steeper than the observations, and with this fit we require the galaxies below about $40\,M_{\odot}\,\mathrm{yr}^{-1}$ to be optically thin to match the FIRAS background measurement[13]. However, given the current uncertainty in the observations and the simplicity of our modeling, this level of agreement is encouraging.

Using our preferred SED, we calculate the redshift distribution of our sub-mm sources (Figure 2). The median redshift is 2.3 and there is a broad distribution in redshift, in agreement with current observational indications.

What does the simulation say about the physical properties of sub-mm sources? We find that the sub-mm galaxies are generally quite massive (several times M_*). At high redshifts, the galaxy mass and star formation rates are strongly correlated, implying that the simulated SCUBA sources are not bright solely due to mergers or bursts. By examining the star formation histories of the galaxies directly, we find that they are often forming stars fairly steadily, although minor bursts do accentuate their brightnesses.

We can also trace the galaxies down to the present day. Their descendants typically have quite old stellar populations, are massive, and are found in dense environments. While the spatial resolution in our simulation does not allow us to distinguish between spirals and ellipticals, all of these indications seem consistent with the notion that SCUBA sources are massive ellipticals in the process of formation.

References

1. Blain, A. W., Smail, I., Ivison, R. J. & Kneib, J.-P. 1999, MNRAS, 302, 632
2. Guiderdoni, B., Hivon, E., Bouchet, F. R. & Maffei, B. 1998, MNRAS, 295, 877
3. Katz, N., Weinberg D.H., & Hernquist, L. 1996, ApJS, 105, 19
4. Davé, R., Dubinski, J., & Hernquist, L. 1997, New Astron, 2, 227
5. Katz, N., Hernquist, L., & Weinberg, D. H. 1999, ApJ, 523, 463
6. Weinberg, D. H., Hernquist, L., & Katz, N. astro-ph/0005340
7. Hernquist L., Katz, N., Weinberg, D.H., & Miralda-Escudé, J. 1996, ApJ, 457, L5
8. Smail, I., Ivison, R., Blain, A., Kneib, J.-P. 2000, this conference
9. Barger, A. J., Cowie, L. L. & Sanders, D. B. 1999, ApJL, 518, L5
10. Blain, A. W., Kneib, J.-P., Ivison, R. J. & Smail, I. 1999, ApJL, 512, L87
11. Chapman, S. C., Scott, D., Borys, C., Fahlman, G. G, 2000, astro-ph/0009067
12. Dunne, L., Eales, S., Edmunds, M., Ivison, R., Alexander, P. & Clements, D. L. 2000, MNRAS, 315, 115
13. Fixsen, D. J., Dwek, E., Mather, J. C., Bennett, C. L. & Shafer, R. A. 1998, ApJ, 508, 123
14. Barger, A. J., Cowie, L. L. & Richards, E. A. 2000, AJ, 119, 2092
15. Smail, I., Ivison, R. J., Owen, F. N., Blain, A. W. & Kneib, J.-P. 2000, ApJ, 528, 612
16. Eales, S., Lilly, S., Webb, T., Dunne, L., Gear, W., Clements, D. & Yun, M. 2000, AJ, 120, 2244

ANDREW W. BLAIN

Institute of Astronomy, Madingley Road, Cambridge, CB3 0HA, UK
E-mail: awb@ast.cam.ac.uk

The results of the first generation of submillimeter (submm)-wave surveys have been published. The opening of this new window on the distant Universe has added considerably to our understanding of the galaxy formation process, by revealing a numerous population of very luminous distant galaxies. Most would have been very difficult to identify using other methods. The potential importance of selection effects, especially those connected with the spectral energy distributions (SEDs) of the detected galaxies, for the interpretation of the results are highlighted.

1   Introduction

From the *IRAS* survey, it was clear that the absorption and re-emission of starlight by interstellar dust in the Milky Way, and of both starlight and accretion energy from AGN by dust in external galaxies, is a very important process. The thermal emission spectrum of dust heated by this radiation, which is observed to peak at a wavelength of order $100\,\mu$m, is redshifted into the submm waveband with a very strong K-correction, making high-redshift submm galaxies unusually easy to detect as compared with their low-redshift counterparts. Between redshifts of about 0.5 and 10 the K-correction almost balances the cosmological dimming of a source with a fixed template SED, and so the flux density received from a galaxy is approximately constant.[11] Although the submm waveband is thus a very attractive window for cosmology,[7] it is technically very challenging to image submm radiation at the faint sensitivity levels – several mJy at $850\,\mu$m – required to detect even an ultraluminous galaxy, defined as possessing a far-infrared(IR) luminosity in excess of $10^{12}\,\mathrm{L_\odot}$, at any redshift $z \geq 0.5$.

Systematic submm-wave cosmology has only been possible since the commissioning of the SCUBA camera at the James Clerk Maxwell Telescope in 1997.[34] A handful of known high-redshift galaxies and AGN were detected earlier using single-pixel detectors,[15,19,37] but blank-field surveys were impossible, due to a combination of relatively low sensitivity and a very small field of view. SCUBA brought 37 and 91 detectors operating in the atmospheric windows at 850 and $450\,\mu$m respectively, providing a 2.5-arcmin field of view, and has been used to good effect to make a range of surveys of the high-redshift Universe.[51,36,2,21,3,57,14] Recently, the MAMBO camera at the IRAM 30-m telescope has also produced deep survey images at 1.25 mm.[5] The existing BIMA, IRAM, Nobeyama and OVRO interferometer arrays are very valuable for making sensitive, high-resolution mm-wave observations, but because of their small fields of view, they are not practical survey instruments. Prior to the debut of SCUBA, the one-dimensional multi-channel 350-μm SHARC camera at the Caltech Submm Observatory (CSO) was used to limit the counts of faint submm galaxies, and the relatively large primary beam of the BIMA array at 2.8 mm was exploited to impose the first limit to the mm-wave counts in a mosaicked image of the Hubble Deep Field.[58] The detection of a very significant intensity of submm/far-IR background radiation from the *COBE* FIRAS and DIRBE datasets was achieved in parallel to the first SCUBA surveys.[47,49,33,25]

The prospects for further instrumental developments are excellent. The lithographic manufacture of large arrays of bolometers is now routinely demonstrated.[12] The BOLOCAM detector array[30] that uses one type of this technology has recently undergone its first engineering tests at the CSO. An alternative technology is exploited in the forthcoming SHARC-II[17] camera for the CSO. Bolometers that exploit superconducting and quantum interference devices rather than simple thermistors are also being developed. These promise increased stability and reduced response time, and would lead to a crucial increase in the degree of multiplexing possible in their readouts, and so to much larger arrays. The future SCUBA-II[35] camera for the JCMT and large-format bolometer cameras for the 50-m LMT[50] are expected to exploit such devices. The SMA[59] interferometer array on Mauna Kea will provide the first sensitive, fully-uv-sampled interferometric images in the submm band, and in the future the 64×12 m ALMA array[60] will provide extremely sensitive, high-resolution observations.

Wide-band mm/submm-wave spectrometers[38] are also being developed, using both arrays of heterodyne detectors[24] and dispersive techniques.[10] These instruments will offer the potential for the direct determination of redshifts for mm/submm-selected galaxies from mm-wave observations of CO lines. There is a natural symbiotic relationship between new panoramic bolometer cameras[30,35] and these spectrographs, which together will be capable of both detecting and obtaining redshifts for large samples of high-redshift dusty galaxies without recourse to either optical, near-IR, or even radio telescopes.

The results and consequences of the first SCUBA surveys have been discussed extensively elsewhere.[8,36,21,43,9,3,44,54] Here the potential selection effects in this new waveband[21,43,11] are described in the context of our knowledge of the SEDs of the detected objects.

2 Submm-wave selection effects

2.1 Redshifts of submm-selected galaxies

In mm/submm-wave surveys there is a unique bias in favor of the detection of high-redshift galaxies at the expense of their low-redshift counterparts. The flux density of a galaxy *with a fixed SED* is expected to be approximately constant over a wide range of redshifts from about 0.5 to 10. The increase in the volume element out to $z \simeq 2$ conspires to bias the selection function to greater redshifts; a demonstration of this effect in currently popular world models as a function of SED is presented elsewhere.[11]

There is broad agreement[44,54] that the objects detected in SCUBA surveys are at redshifts $z \simeq 3 \pm 2$. All three SCUBA galaxies with redshifts confirmed using CO spectroscopy[26,27,55] are at $z > 1$. Others for which multi-waveband data is available[36,18,52,21,29] have no potential low-redshift counterparts. Four of the first galaxies from the CUDSS survey[22,43] were identified with galaxies at $z < 1$; however, more extensive results[21,44] show a smaller fraction of potential low-redshift identifications. Note that two initially plausible low-redshift identifications of galaxies in the SCUBA Lens Survey[51,53] were subsequently revised upwards in the light of additional data.[52] New results from this meeting[57,13] are consistent with a distant redshift distribution.

Despite their potential for detecting high-redshift galaxies, submm surveys detect galaxies at restframe wavelengths considerably longwards of the peak of their SEDs. It is thus possible that the form of this SED can affect the interpretation of the results of SCUBA surveys.[21,11] The main effect would be to overestimate the bolometric luminosity associated with a submm-selected galaxy if its dust temperature was overestimated.

2.2 SEDs of submm-selected galaxies

Potential selection effects in submm-wave surveys are controlled by the SEDs of the detected galaxies. The SEDs of individual dusty galaxies certainly vary from galaxy to galaxy, and could vary systematically with luminosity and redshift. Here and in previous work connected with the SCUBA Lens Survey,[8,9,11] three parameters are used to describe the SED; a dust temperature T in a standard Planck function B_ν, an index β in a dust emissivity function ν^β and a spectral index α to describe the mid-IR SED, $f_\nu \propto \nu^\alpha$.

Dust SEDs are steeper than blackbody spectra in the Rayleigh–Jeans (RJ) regime. If the SED is represented by the function $\nu^\beta B_\nu$, then the RJ spectral index is $-2-\beta$. Some authors[4,45] represent the SED by a greybody of the form $(1 - \exp^{-K\nu^\beta})B_\nu$, explicitly taking account of the increase in optical depth with increasing frequency. In this case, the RJ spectral index is also $-2 - \beta$, but the SEDs differ near their peaks. Temperatures fitted using a greybody SED are typically about 20% greater than those derived from the $\nu^\beta B_\nu$ SED.

The mid-IR α term is introduced because dust SEDs do not fall exponentially like blackbodies in the Wien regime, due to the contribution from additional hot dust components in the interstellar medium. At frequencies above which the gradient of $\nu^\beta B_\nu$ is steeper than that of the power law ν^α, the SED is represented by this power law, thus incorporating the observed mid-IR properties of galaxy SEDs; see Fig. 1. The radio emission associated with dusty galaxies is represented by an additional power-law function, normalized to fit the low-redshift far-IR–radio correlation.[16]

These three simple parameters can provide an adequate and appropriate description of the SEDs of distant dusty galaxies from the mm to mid-IR waveband,[6] which are constrained only by a handful of broad-band photometric measurements with little or no spatial resolution. When sub-arcsecond spatially-resolved images from ALMA are available, it will be important to investigate more details of the mm to near-IR SEDs, using radiative transfer models, including full details of the complex geometry, the different physical components and conditions in a merging, star-forming galaxy, and of a potential point-like nuclear heat source.[23,32] However, few of the parameters required by such models can be constrained at present, and so we prefer to use the simple three-parameter SED. Based on fits to the form of evolution of low-redshift 60-μm $IRAS$ galaxies and more distant galaxies selected at 175 μm, described elsewhere,[8] values of $T = 38\,\mathrm{K}$, $\beta = 1.5$ and $\alpha = -1.7$ were chosen; the associated SED is shown by the dashed line in Fig. 1.

Information about the SEDs of the high-redshift luminous galaxies detected in submm surveys can be obtained by several different routes.

First, submm-wave observations of a representative sample of galaxies from the $IRAS$ catalog can be made.[20] However, being at low redshifts ($z \leq 0.1$), the properties of these galaxies have rather little overlap with those of the more distant, typically much more luminous galaxies found in deep submm surveys. The results indicate that there is a weak trend for the dust temperature to increase with increasing luminosity, and that $T = 36 \pm 5\,\mathrm{K}$ and $\beta = 1.3 \pm 0.2$ are reasonable typical values; see the dotted SED shown in Fig. 1. Independent results for the typical SEDs of $IRAS$ galaxies, which yields a similar generic spectrum, are shown by the solid line in Fig. 1.[31]

Secondly, color information for high-redshift submm-selected dusty galaxies can be obtained from multiband submm/mm/IR observations and combined with independent redshift information to derive T and β.[40,41] The few temperatures available are in the range 40-50 K, consistent with the SEDs described above. The relevant data are plotted in Fig. 1.[11,48] Note that because dust SEDs are thermal, it is impossible to distinguish *a priori* between the effects of an increase in temperature or a reduction in redshift.

Thirdly, the color technique can be used to measure SEDs for high-redshift galaxies with known redshifts, but a less uniform selection criterion, for example the most luminous *IRAS* galaxies, high-redshift radio galaxies, optically-selected AGN and gravitational lenses.[11,48] The ranges of luminosity and redshift that are appropriate to the SCUBA galaxies are sampled using this approach; however, the surveyed objects are diverse, and perhaps extreme and unrepresentative. For example, the dust temperature inferred from observations of the most luminous lensed quasar[42,6] is about three times greater than the typical temperature inferred from *IRAS*, *ISO* and SCUBA data.

In general, dust temperatures in high-redshift AGN seem to be greater than in *IRAS* and SCUBA galaxies. This is probably due both to the selection effect against detecting hot objects in SCUBA surveys, and to higher intrinsic temperatures in the most luminous objects.

2.3 Pointed observations

There has been much concerted effort to detect statistical samples of known high-redshift galaxies, such as the Lyman-break galaxies,[13] classical radio galaxies,[1] quasars[4,39] and gravitational lenses using mm/submm instruments. Before SCUBA, this technique provided the only opportunity to study the high-redshift mm/submm Universe.[15,19,37] Today, it provides a direct opportunity to connect the results obtained from deep submm surveys to more developed fields of observational cosmology carried out in other wavebands.

The selection functions at work in these studies favor the detection of the highest redshift members of flux-limited samples of objects selected in another waveband, because of the familiar mm/submm K-correction. The ratio between the submm flux density of a galaxy and that measured in any other waveband, with the possible exception of soft X-ray observations of highly-enshrouded gas-rich AGN, is expected to be a strongly increasing function of redshift. For example, the submm–radio/optical flux density ratio of a galaxy with a reasonable radio or optical SED, $f_\nu \propto \nu^{-1}$, and a submm SED, $f_\nu \propto \nu^{3.5}$, should increase with redshift z as $(1 + z)^\gamma$, where $\gamma \simeq 4.5$. The SCUBA detection of a $z = 5.8$ quasar discovered in the Sloan Digital Sky Survey reported at this meeting, despite the typical lack of detections of $z \simeq 3$ quasars,[39] provides a possible example. It is exciting that the growing number of very high-redshift objects detected in other wavebands could be relatively

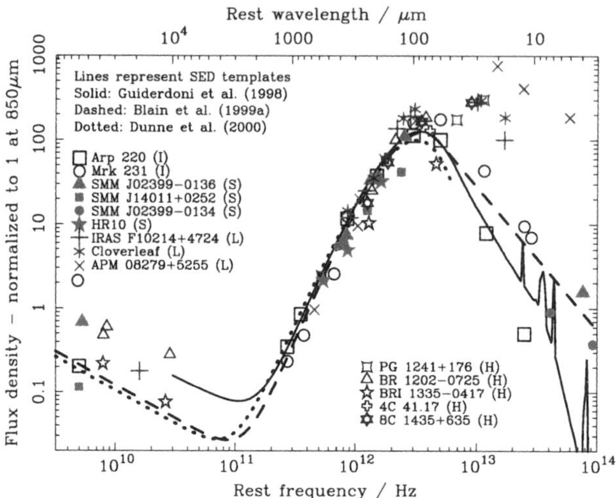

Figure 1. SEDs of a wide range of dusty galaxies with known redshifts, including known lensed sources (L), high-redshift AGN (H), low-redshift *IRAS* galaxies (I) and submm-selected galaxies (S). Three spectral templates are also shown. Only those objects with redshifts and both mid-IR and submm data are plotted. Both the re-normalized data and the templates are very similar at wavelengths longer than 60 μm.

easy to detect in pointed mm/submm observations, using existing telescopes like the JCMT and CSO, forthcoming telescopes like the SMA and LMT, and ultimately at great sensitivity and resolution using ALMA. The intergalactic medium is always transparent to mm/submm radiation, and so this type of observation could be very important for tracing the process of re-ionization in the years ahead.

It should be possible to study the intrinsic evolution of the dust emission properties of a well-defined sample of galaxies by making mm/submm observations of a sub-sample that have been selected carefully elsewhere to have a fixed luminosity as a function of redshift.[1] For example, while the fraction of powerful radio galaxies drawn from flux-limited samples that are detected by SCUBA is observed to increase with redshift, consistent with the selection effect above, a sub-sample with matched radio luminosities still displays a statistically significant increase in dust luminosity with redshift, indicating that powerful radio galaxies are systematically more luminous in the submm than the radio waveband at higher redshifts.[1]

3 Summary: future observations of very early galaxies

Selection effects in deep mm/submm-wave surveys generally favor the detection of the most distant objects. For a fixed bolometric luminosity, dusty galaxies with colder dust temperatures are more likely to be detected. The detection of galaxies at the highest redshifts in the mm/submm waveband requires only that they are luminous and contain dust. Because absorption along the line of sight is not significant in the mm/submm waveband, dust generated and heated in the very first galaxies, prior to the epoch of re-ionization, could potentially be detected in the submm waveband.

Acknowledgments

I thank the Raymond and Beverly Sackler Foundation for support at the IoA, UMass/INAOE for support at the meeting, the CfA for organizing 'The First Generation of Cosmic Structures', for which some of this material was prepared, and Jan and Rick Caldwell for hospitality prior to both meetings.

References

1. E.N. Archibald, *et al*, *MNRAS*, submitted, astro-ph/0002083 (2000).
2. A.J. Barger, L.L. Cowie, *et al*, *Nature* **394**, 247 (1998).
3. A.J. Barger, L.L. Cowie, and D.B. Sanders, *ApJ* **518**, L5 (1999).
4. D.J. Benford, P. Cox, *et al*, *ApJ* **518**, L65 (1999).
5. F. Bertoldi, *et al*, *A&A*, in press, astro-ph/0006094 (2000).
6. A.W. Blain, *MNRAS* **304**, 699 (1999).
7. A.W. Blain, and M.S. Longair, *MNRAS* **264**, 509 (1993).
8. A.W. Blain, I. Smail, R.J. Ivison, *et al*, *MNRAS* **302**, 632 (1999).
9. A.W. Blain, A. Jameson, *et al*, *MNRAS* **309**, 715 (2000).
10. A.W. Blain, D.T. Frayer, J.J. Bock, *et al*, *MNRAS* **313**, 559 (2000).
11. A.W. Blain, *et al*, ASP Conf. Ser. **193**, 425 (2000).
12. J.J. Bock, *et al*, Proc. SPIE **3357**, 297 (1998).
13. S.C. Chapman, *et al*, *MNRAS*, in press, astro-ph/9909092 (2000).
14. S.C. Chapman, *et al*, *MNRAS*, submitted (2001).
15. D.L. Clements, *et al*, *MNRAS* **256**, P35 (1992).
16. J.J Condon, *ARA&A* **30**, 575 (1992).
17. C.D. Dowell, S.H. Moseley Jr., *et al*, ASP Conf. Ser., in press (2000).
18. D. Downes, *et al*, *A&A* **347**, 809 (1999).
19. J.S. Dunlop, D.H. Hughes, *et al*, *Nature* **370**, 437 (1994).
20. L. Dunne, *et al*, *MNRAS* **315**, 115 (2000).

21. S.A. Eales, *et al*, *MNRAS*, in press (2000).

22. S.A. Eales, *et al*, *ApJ* **515**, 518 (1999).

23. A. Efstathiou, M. Rowan-Robinson, *et al*, *MNRAS* **313**, 734 (2000)

24. N. Erickson, this volume (2001).

25. D.P. Finkbeiner, M. Davis, *et al*, *ApJ*, in press, astro-ph/0004175 (2000).

26. D.T. Frayer, R.J. Ivison, N.Z. Scoville, *et al.*, *ApJ* **506**, L7 (1998).

27. D.T. Frayer, R.J. Ivison, N.Z. Scoville, *et al.*, *ApJ* **514**, L13 (1999).

28. D.T. Frayer, I. Smail, *et al*, *AJ*, in press, astro-ph/0005239 (2000).

29. W.K. Gear, *et al*, *MNRAS*, in press, astro-ph/0007054 (2000).

30. J. Glenn, *et al*, *Proc. SPIE* **3357**, 326 (1998).

31. B. Guiderdoni, E. Hivon, F.R. Bouchet, *et al*, *MNRAS* **295**, 877 (1998).

32. B. Guiderdoni, and J.E.G. Devriendt, astro-ph/9911162, (2000).

33. M.G. Hauser, *et al*, *ApJ* **508**, 25 (1998).

34. W.S. Holland, *et al*, *MNRAS* **303**, 659 (1998).

35. http://www.jach.hawaii.edu/JACpublic/JCMT/scuba/scuba2.

36. D. Hughes, *et al*, *Nature* **394**, 241 (1998).

37. K.G. Isaak, R.G. McMahon, *et al*, *MNRAS* **269**, L28 (1994).

38. K.G. Isaak, A. Harris, and J. Zmuidzinas, *BAAS* **191**, 0911 (1997).

39. K.G. Isaak, *et al*, this volume (2001).

40. R.J. Ivison, I. Smail, J.-F. Le Borgne, *et al*, *MNRAS* **298**, 583 (1998).

41. R.J. Ivison, I. Smail, A.J. Barger, *et al*, *MNRAS* **315**, 209 (2000).

42. G.F. Lewis, S.C. Chapman, R.A. Ibata, *et al*, *ApJ* **505**, L1 (1998).

43. S.J. Lilly, *et al*, *ApJ* **518**, 641 (1999).

44. S.J. Lilly, *et al*, this volume (2001).

45. U. Lisenfeld, K.G. Isaak, and R.E. Hills, *MNRAS* **312**, 433 (2000).

46. T.G. Phillips, ESA SP-401, 223 (1997).

47. J.-L. Puget, *et al*, *A&A* **308**, L5 (1996).

48. M. Rowan-Robinson, *MNRAS* **316**, 885 (2000).

49. D.J. Schlegel, D.P. Finkbeiner and M. Davis, *ApJ* **500**, 525 (1998).

50. F.P. Schloerb, *Proc. IAU* **170**, 221 (1997).

51. I. Smail, R.J. Ivison and A.W. Blain, *ApJ* **490**, L5 (1997).

52. I. Smail, R.J. Ivison, J.-P. Kneib, *et al*, *MNRAS* **308**, 1061 (1999).

53. I. Smail, R.J. Ivison, A.W. Blain, *et al*, *ApJ* **507**, L21 (1998).

54. I. Smail, R.J. Ivison, *et al*, this volume, astro-ph/0008237, (2001).

55. G. Soucail, J.-P, Kneib, J. Bézecourt, *et al*, *A&A* **343**, L70 (1999).

56. N. Trentham, A.W. Blain and J. Goldader, *MNRAS* **305**, 61 (1999).

57. P. van der Werf, *et al*, this volume (2001).

58. D.J. Wilner and M.C.H. Wright, *ApJ* **488**, L67 (1997).

59. D.J. Wilner, *et al*, this volume (2001).

60. A. Wootten, ed., *Science with ALMA, ASP Conf. Ser.*, in press (2000).

PARTICIPANTS

Name	Affiliation	Email Address
Barger, Amy	University of Hawaii	barger@ifa.hawaii.edu
Barvainis, Richard	NSF	rbarvai@nsf.gov
Blain, Andrew	IoA, University of Cambridge	awb@ast.cam.ac.uk
Borys, Colin	U. British Columbia	borys@physics.ubc.ca
Bouché, Nicolas	University of Massachusetts	bouche@nova.astro.umass.edu
Carilli, Chris	NRAO	ccarilli@nrao.edu
Carrasco, Luis	INAOE	carrasco@inaoep.mx
Chapman, Scott	Carnegie Observatories	schapman@ociw.edu
Choi, Junhwan	University of Massachusetts	jhchoi@nova.astro.umass.edu
Ciardi, Benedetta	Universitá degli Studi di Firenze	ciardi@arcetri.astro.it
Clements, David	Cardiff University	d.clements@astro.cf.ac.uk
Combes, Francoise	Observatoire de Paris	francoise.combes@obspm.fr
Devlin, Mark	University of Pennsylvania	devlin@physics.upenn.edu
DeVries, Chris	University of Massachusetts	devries@astro.umass.edu
Dicker, Simon	University of Pennsylvania	sdicker@upenn5.hep.upenn.edu
Dunlop, James	University of Edinburgh	jsd@roe.ac.uk
Fardal, Mark	University of Massachusetts	fardal@weka.astro.umass.edu
Frayer, David	Caltech	dtf@astro.caltech.edu
Gaztañaga, Enrique	INAOE	gazta@inaoep.mx
Gundersen, Joshua	Princeton University	gunder@pupgg.princeton.edu
Gutierrez, Fidel	INAOE	fgr@inaoep.mx
Hooper, Eric	Harvard-Smithsonian CfA	ehooper@cfa.harvard.edu
Hughes, David	INAOE	dhughes@inaoep.mx
Iono, Daisuke	University of Massachusetts	diono@nova.astro.umass.edu
Isaak, Kate	Cavendish / U. Cambridge	isaak@mrao.cam.ac.uk
Ivison, Rob	University College London	rji@star.ucl.ac.uk
Katz, Neal	University of Massachusetts	nsk@kaka.astro.umass.edu
Klein, Jeff	University of Pennsylvania	jklein@physics.upenn.edu
Knudsen, Kirsten Kraiberg	Leiden Observatory	kraiberg@strw.leidenuniv.nl
Kraan-Korteweg, Renee	U. Guanajuato	kraan@astro.ugto.mx
Lilly, Simon	University of Toronto	lilly@astro.utoronto.ca
Linder, Suzanne	INAOE	slinder@inaoep.mx
Lowenthal, James	University of Massachusetts	james@astro.umass.edu
McMahon, Richard	Institute of Astronomy,	rgm@ast.cam.ac.uk
Menendez, John	Composite Optics, Inc.	jmenendez@coi-world.com
Miller, Neal	NRAO/NMSU	nmiller@aoc.nrao.edu
Neininger, Nikolaus	Radioastr. Inst. Univ. Bonn	nneini@astro.uni-bonn.de

Ngeow, Choong	University of Massachusetts	ngeow@nova.astro.umass.edu
Olmi, Luca	LMT Project/UMass	olmi@lmtgtm.org
Robson, Ian	JAC	eir@jach.hawaii.edu
Rosa González, Daniel	INAOE	danrosa@inaoep.mx
Sanders, David	University of Hawaii	sanders@ifa.hawaii.edu
Schloerb, Pete	University of Massachusetts	schloerb@astro.umass.edu
Scott, Douglas	University of British Columbia	dscott@astro.ubc.ca
Scoville, Nick	Caltech	nzs@astro.caltech.edu
Shaver, Peter	European Southern Observatory	pshaver@eso.org
Smail, Ian	University of Durham	ian.smail@durham.ac.uk
Snell, Ronald	University of Massachusetts	snell@fcraol.astro.umass.edu
Taylor, Chris	UMass / FCRAO	chris@fcrao1.astro.umass.edu
Terlevich, Elena	INAOE	eterlevi@inaoep.mx
Terlevich, Roberto	IoA; INAOE	rjt@ast.cam.ac.uk
van der Werf, Paul	Leiden Observatory	pvdwerf@strw.leidenuniv.nl
Webb, Tracy	University of Toronto	webb@astro.utoronto.ca
Wilner, David	Harvard-Smithsonian CfA	dwilner@cfa.harvard.edu
Wolf, Marsha	University of Texas at Austin	mwolf@astro.as.utexas.edu
Young, Judith	UMass and FCRAO	young@astro.umass.edu
Yun, Min	University of Massachusetts	myun@astro.umass.edu
Zivkov, Vesna	University of Massachusetts	vzivkov@nova.astro.umass.edu

ERRATA

AUTHOR INDEX

AUTHOR INDEX